THE QUANTUM EXODUS

THE QUANTUM EXODUS

Jewish Fugitives, the Atomic Bomb, and the Holocaust

by

Gordon Fraser

OXFORD
UNIVERSITY PRESS

Great Clarendon Street, Oxford OX2 6DP

Oxford University Press is a department of the University of Oxford.
It furthers the University's objective of excellence in research, scholarship,
and education by publishing worldwide in

Oxford New York

Auckland Cape Town Dar es Salaam Hong Kong Karachi
Kuala Lumpur Madrid Melbourne Mexico City Nairobi
New Delhi Shanghai Taipei Toronto

With offices in

Argentina Austria Brazil Chile Czech Republic France Greece
Guatemala Hungary Italy Japan Poland Portugal Singapore
South Korea Switzerland Thailand Turkey Ukraine Vietnam

Oxford is a registered trade mark of Oxford University Press
in the UK and in certain other countries

Published in the United States
by Oxford University Press Inc., New York

First published 2012

British Library Cataloguing in Publication Data
Data available

Library of Congress Cataloging in Publication Data
Library of Congress Control Number: 2011944061

Typeset in Spectrum MT by Cenveo, Bangalore, India
Printed and bound by
CPI Group (UK) Ltd, Croydon, CR0 4YY

ISBN 978–0–19–959215–9

1 3 5 7 9 8 6 4 2

This book is dedicated to my father, Jack Fasht, 1912 –

Contents

List of illustrations

Neutrons and Nazis

It had become dark in physics, from the top downwards. With the massive infiltration of the Jews into important posts in universities and academies, the observation of Nature—the basis for all natural science—had been forgotten. Instead, knowledge was supposed to be based on human imagination. . . . The most obvious example of this damaging influence of the Jews on science was provided by Herr Einstein.

Philipp Lenard, winner of the 1905 Nobel Prize for Physics, writing in the *Völkischer Beobachter*, 15 May 1933.

After such prejudice, it was no accident that the Atomic Bomb and the Holocaust emerged simultaneously from the turmoil of the twentieth century. In its first thirty tumultuous years, many political and demographic threads had become tightly knotted. Like a puppet whose strings have become entangled, an ill-timed jerk brought unexpected results.

In 1911, as a belligerent world lurched towards war, these various threads were still largely separate. Nations vied with each other in a hectic arms race, each boasting of a perceived superiority. In all this swashbuckling defiance, Great Britain still considered itself the world's leading sea power. To maintain this claim, plans had been drawn for a new breed of battleship, the 'super-dreadnought', bigger and superior in firepower, armour, and speed to anything then afloat. Super-dreadnoughts also brought other innovations. Instead of the traditional coal, of which Britain then had an enormous supply, new battleships would instead be fuelled by imported oil, presaging a new era in geopolitics. They were also to be armed with 15-inch calibre guns, capable of hurling a one-tonne shell over a distance of 30 kilometres.

That year a 40-year-old New Zealand scientist working at the University of Manchester made a startling discovery and struggled to find an apt analogy: 'It was as though you had fired a 15-inch naval shell at a sheet of tissue paper and it had come back and hit you'. Ernest Rutherford was speaking figuratively, of course. Had he really been hit by a shell from a super-dreadnought, he would not have had much time to be surprised. But Rutherford had just discovered the atomic nucleus, hidden deep in the heart of the already infinitesimal atom. It was a new milestone in the quest to understand the innermost structure of matter, and was even more significant than the work for which Rutherford had already won a Nobel Prize several years earlier. His nuclear revelation would be more consequential than any battleship.

Nevertheless, in 1911 other people were more concerned about politics and the imminence of war rather than the science of invisible atoms and their even more invisible nuclei. Rutherford's quote reflects as much the aggressive belligerence of the time as it does about the workings of the atom. In the lead-up to the conflict that would become the First World War, the super-dreadnoughts with their 15-inch guns were still the ultimate in naval firepower.

Across the North Sea, Germany was also building battleships. After the outbreak of war in 1914, German and British ships skirmished sporadically until 1916, when the German High Seas Fleet boldly steamed out to challenge the super-dreadnoughts and other ships of the British Grand Fleet at the Battle of Jutland. One of the biggest naval battles in history, it was nevertheless inconclusive. The confused and disorganized British lost more ships, but the German fleet remained bottled-up in port for the remainder of the war, its frustrated commanders seething to underline their new claim to naval superiority. After the sudden 1918 armistice, the German fleet ignominiously crossed the North Sea to be interned at the British naval base at Scapa Flow.

When the First World War began, three years after Rutherford's nuclear realization, science and scientists on both sides were mobilized for the war effort. Talented German chemists (several went on to earn Nobel prizes) developed new weapons of mass destruction using poison gas. On the Allied side, Rutherford had been given other responsibilities. In 1912, the focus of world attention had been temporarily diverted from the inevitability of war by the sinking of the *Titanic* after it struck an iceberg on its maiden voyage. The *Titanic* had been equipped with radio, but this could not in itself detect icebergs. Could sound waves be used to detect such underwater obstacles and avert future disasters? After the outbreak of the war and the emergence of submarines as a threat to shipping, the goal of this sonic effort shifted. As well as sinking merchant shipping, invisible submarines also posed a major threat to the huge super-dreadnoughts, which spent much of their time in the First World War confined to their bases, reluctant to venture onto the open seas to provide a ready target for torpedoes. Playing a major role in British anti-submarine development work was Rutherford.[1]

With world attention focused on the carnage of that war, atomic science was an obscure corner of academia. But Rutherford's discovery of the atomic nucleus had nevertheless been a huge stimulus: when they found time, researchers turned their attention from the atom as a whole to what went on deep in its centre. Continually in the forefront of the effort was Rutherford. When not developing underwater sonics, he persevered dutifully with his nuclear investigations, firing the atomic equivalent of 15-inch shells at unsuspecting nuclear targets. Nuclei are extremely small, and the chance of hitting one was like target practice in thick fog: most shots missed their targets entirely. But among all the subatomic artillery shells emerging intact, Rutherford spotted a few lighter fragments. A few of Rutherford's fat projectiles had broken up after slamming into nuclear targets. The ancient alchemists had dreamed of transforming base metals into gold; Rutherford's artillery had instead transmuted nitrogen nuclei into those of oxygen. While he was making this discovery, in October 1918 a German submarine attempting to enter the British Scapa Flow naval base was detected by the new anti-submarine equipment and promptly destroyed by mines triggered from the shore. After arriving late when unexpectedly summoned to another meeting on anti-submarine measures, Rutherford excused himself: 'If, as I have reason to believe, I have disintegrated the nucleus of the atom, this is of greater significance than the war'.[2]

The fragments seen by Rutherford were protons, the bricks from which nuclei are built. Protons added a new scale to the map of Nature. Bill Bryson's magnificent book *A Short History of Nearly Everything* begins 'No matter how hard you try you will never be able to grasp just how tiny ... is a proton. ... A little dib of ink like the dot on this 'i' can hold ... 500,000,000,000 of them, rather more than the number of seconds [in] half a million years'.[3] These ungraspable protons conveniently label different kinds of nuclei. The simplest, those of hydrogen, have a lone proton: the heaviest, those of uranium, have 92.

When research scientists finally prise open a locked door and eagerly peer inside, they do not always discover dazzling treasure. Often they are disappointed to see a short, dimly lit corridor leading to yet another door. Rutherford's proton revelation was like this. But there was nevertheless enough illumination in the corridor for him to suspect that nuclei had to contain more than just his newly discovered protons. A lot more. A uranium nucleus is some two and a half times heavier than its 92 protons. To explain so much extra material, Rutherford realized that something else lurked inside the nucleus. He took an option on a new name—neutron—and told his research team to start looking. This door was difficult to unlock, and by the time one of Rutherford's team finally succeeded, the world had changed.

When Rutherford first began talking about neutrons, an ex-soldier called Adolf Hitler was beginning to make a name for himself in and around Munich, addressing meetings of an obscure movement, the *Nationalsozialistische Deutsche Arbeiterpartei* (the Party of Nationalist Socialist German Workers, or NSDAP—abbreviated in English to 'Nazi'). By the time the neutron was finally discovered in 1932, the Nazi movement had progressed from the fringes of Bavarian politics to become the dominant force in Germany.

In that time, science too had advanced. The focus had shifted from the already infinitesimal atom to an atomic nucleus many thousands of times smaller. Scientists knew what held atoms together—conventional explosives could release this stored energy—but whatever held the tightly packed nucleus together was a mystery. Einstein's new theory of relativity had been hinting that energy could be transformed in new ways, and the detailed accounting of protons and neutrons suggested that immense energy was somehow concentrated inside nuclei. Could this hidden nuclear energy be tapped?

Sometimes scientific experiments confirm what is already known, or extend this knowledge. At other times their results can be unexpected, even surprising, and conventional wisdom has to be called in for an overhaul. Thus it was for the interior of atoms. The laws which governed this realm appeared to be written in some arcane code. Most of the scientists who could decipher these quantum hieroglyphs were either German, or came from countries which Nazi Germany was soon to annex. Curiously, many of them also happened to have Jewish backgrounds.

When the Nazis came to power in 1933, their initial objective was not to get rid of Jews. That would come later. A more pressing aim was a new national enlightenment, to refine German culture and purge it of perceived pollution by Jews. Culture does not grow on trees: it has to be imparted. One of the first Nazi laws targeted the German machinery of learning: Jewish professors and teachers were dismissed from the nation's fine universities. All faculties—humanities, law, music, ... were affected, but the impact was particularly severe on atomic science, which had attracted an unusually high concentration of Jewish talent. German science had become a symbol of the nation's sovereignty: German was by default the premier language of international science. But in striving to cleanse its culture, Nazi Germany undermined one valuable inheritance. The structure did not fall down—indeed parts of it survived unscathed—but its profile was reshaped.

On the first page of his magisterial book *Nazi Germany and the Jews,*[4] Saul Friedländer begins 'The exodus from Germany of Jewish ... artists and intellectuals began in the early months of 1933, almost immediately after Adolf Hitler's accession to power on 30 January'. But Friedländer overlooks one influential scientist who had already jumped Hitler's gun—Albert Einstein, the figurehead of twentieth-century science, and one of the first Jewish scientists of world rank since the twelfth century. Einstein led this exodus by example.

Jews had a long tradition of study: religious minutiae had busied generations of students. But in nineteenth-century Germany, assimilated Jews had begun to look to new intellectual horizons. With Einstein as a radiant role-model, their talents later turned to new challenges. These scientists helped fashion a mysterious new formalism, quantum mechanics, the calculus of the intangible domain inside the atom. After losing their jobs in 1933, if they wanted to continue being research scientists, they had to go into exile. Emigration is never easy, but these quantum priests were known, even revered, and they were welcomed with open arms.

Just as Hitler came to power, Rutherford's long-anticipated neutron finally showed itself. Today's discovery quickly becomes tomorrow's research tool, and so it was with neutrons. A neutron trigger could upset a delicately balanced nucleus, releasing stored energy. Some of the scientists who made these discoveries happened to be those targeted by the new Nazi laws. As these quantum exiles scattered, they realized that they held the key to a new weapon of unimaginable power. For them, it was not a great intellectual leap, so they were convinced that their gentile counterparts in Nazi Germany had come to the same conclusion. By 1939, the Nazi attitude towards the Jews had hardened. What had begun as prejudice had escalated into outright persecution, and would soon become annihilation. After having witnessed what the Nazis were prepared to do, the exiles were afraid. When the world erupted again into war, they had to get to the Atomic Bomb first.

Unknown to the exiles, and to the world, the Nazis had chosen not to follow this road. They had another objective, equally inconceivable, and to them more important. In Germany, and later in Austria, they had been able to make Jews emigrate by making life difficult for them. As German armies pushed eastwards across Poland and the Soviet Union, they overran the heartland of world Jewry. Populations numbered in millions were far too large for mass emigration. The 'final solution' to this dilemma was genocide, what was to become known as the Holocaust. To succeed, this needed logistics and resources on a massive scale. So did the Atomic Bomb. Because of their sheer size, the Nazis could not have engineered both of them simultaneously. Each justified by its own crude bluntness, two dreadful projects forged ahead in parallel, and each in turn became reality. On two counts, what had been unimaginable no longer was.

So frightened of so few

The total world Jewish population is about 13 million, or about one person in 500.[5] The recent history of this tiny minority is dominated by the Holocaust and by the establishment of the state of Israel. But who and what were the Jews before this? Why did they come to be resented in Germany? How could Hitler be so frightened of so few, making them his primary enemy, dedicating a vast effort to their annihilation, even when the very defences of his nation were crumbling?

The gas argon makes up about one per cent of the air we breathe. This is not very much, but it is nevertheless about twenty-five times more plentiful in the atmosphere than carbon dioxide, without which life as we know it would not exist. Unlike this vital CO_2, argon is stand-offish: its very name comes from the Greek 'inert'. The word later came to be applied to a whole family of chemical elements, the 'inert gases', because of their extreme reluctance to participate in chemistry. Sometimes called 'rare gases', because the others are even scarcer than argon, their only role is to fill an array of slots in the Periodic Table of the elements.

The Periodic Table is also the title of a book by Primo Levi, a gifted industrial chemist whose scientific knowledge underpins this collection of short stories, each depicting the special role that a particular chemical element appeared to play in his life. Born in 1919 into a cultured Jewish family in Turin, Levi's career was initially handicapped when racial laws in Fascist Italy stigmatized him as a Jew. Unable to find a wartime job, he joined an Italian resistance movement. With a complete lack of military training, he was soon arrested by the German forces occupying Northern Italy, and confessed to being Jewish to avoid being shot as a partisan. This was the first example of his extraordinary ability to survive, which took him through the horror of Auschwitz and its grim aftermath. His train carried away more than six hundred Italian Jews. Most were gassed within two days of arrival. Of the 125 selected as fit enough to be worked to death, he was one of just three survivors.[6]

The opening chapter of Levi's book hesitantly compares the almost invisible trace of argon in the atmosphere to the Jews in humanity. Levi's Jew–argon analogy is perceptive, but as he himself acknowledges, it is not altogether valid. Colourless and chemically inert, argon hides in the jostle of the atmospheric crowd. It does not have to do anything to justify itself or ensure its continued existence. It is simply there, the smug occupant of a place guaranteed by the quantum laws that determine how atoms can be made up. In stark contrast to argon, inertness is not a typically Jewish trait. As Levi points out, his ancestors 'were—or had to be—quite active in order to earn a living'. Jews striving to survive could not afford the luxury of being inert, and often had to be spectacularly busy even to emulate the modest achievements of others. Jews are also about five times rarer than argon.

When Fascist racial laws restricted the traditional freedom of Italian Jews, Levi began to think of himself and his family in a new role. Instead of being hidden inside a seemingly homogeneous mix, they had become instead a visible impurity. For a chemist like Levi, an impurity can play very different roles. Some substances—such as silicon wafers for modern microelectronics—have to be extremely pure to be effective. Purity can become an ideal: the Nazis saw racial integrity as a prime objective, using misguided logic based on warped ideas of genetics. But on the other hand some 'impurities' can enhance the properties of their host. Traces of carbon turn dull iron into fine steel. Another example of a beneficial impurity is a catalyst, which can radically change chemical dynamics. Tiny amounts of rare metals such as platinum or palladium absorb carbon monoxide from vehicle exhaust fumes. Enzymes are

Nature's own catalysts, for example breaking up complex foods into more digestible material.

Throughout history, it has been the fate of Jews to lurch periodically from being seen as troublesome impurities, to being nurtured as valuable catalysts. Each such outcome generates its own backlash: persecution generates pity, while success breeds jealousy and resentment. Jewish history thus appears as an unpredictable rollercoaster ride, continually swinging between tragedy and triumph. Nevertheless, Jews have resolutely clung to their identity for some four thousand years, an identity which itself has been moulded by a turbulent history which spans much of that of civilization itself. Jewish history dovetails with that of empires which have disappeared over the horizon of antiquity—Egypt, Persia, Babylon, and Rome. After floundering on the floodtides of these mighty imperia, the Jews have nevertheless survived them.

This protracted chronicle appears to break into two contrasting periods, each of two thousand years, hinged at the Roman occupation of the Jewish homeland of Judea, the era when Jesus Christ lived and preached.[7] The end of this first period was symbolically marked by the destruction of the Temple in Jerusalem in 70AD and the massacre of the city's population,[8] commemorated in the Jewish calendar by a fast on the ninth day of the month of Av (*Tish b'Av*). Thereafter, the citizens of Judea dispersed, while Judaism changed from the creed of an obscure clan into a template for world religions. The soldiers of Judea had been fierce warriors, but with this diaspora, the Jews lost the military capability to resist any outright attack, and became vulnerable.

On such a fragile perch, one route to acceptance was to seek ways of ingratiating themselves to their hosts. Many oriental countries relied on Jewish doctors. These healers and pharmacists were among the first Jewish scientists. The greatest was the polymath Rabbi Moses ben Maimon (1135–1204), also known under the Greek name of Maimonides, or in Hebrew by the acronym *Rambam*. A contemporary in Moorish Spain of the Islamic philosopher known in the West as Averroës, Maimonides helped reconcile the classical philosophy of Ancient Greece, largely forgotten, with that of Islam and Judaism. His *Guide for the Perplexed*, written in Arabic, explored and explained the relationship between religion and philosophy, and influenced thinkers such as Thomas Aquinas. For his own religion, Maimonides' extensive commentary on the intricate canon of Talmudic law is still consulted.

In medicine, he wrote standard textbooks on diagnosis, lifestyle, and drugs, as well as specialized works on diet, poisons, aphrodisiacs, asthma, and haemorrhoids.

The Jews had flourished in a relatively benevolent Islamic regime in Moorish Spain. In the twelfth century, this was suddenly overthrown by a wave of religious fundamentalism from North Africa, the Almohades. Less tolerant of other religions and even of other Moslems, viewed as decadent, the Almohades drove many Jews into a fresh exile, in the case of Maimonides first to Morocco and then Egypt, where he became court physician to the Sultan Saladin. His lifetime contributions to Jewish culture, and to learning in general, made Maimonides a conspicuous and highly regarded ambassador for his people, and a role-model for future generations. He became the figurehead of a new Jewish ideal, which would remain for some seven hundred years until his memory and achievements were rivalled, perhaps even eclipsed, by another great Jewish scientist forced to emigrate from his homeland—Albert Einstein.

In Germany

In Europe, continual power struggles became conflated with fresh rivalries between Catholicism and the new force of the Protestant Reformation. In the aftermath of one of the most destructive of these conflicts, the Thirty Years War (1618–48), Brandenburg's modest capital, Berlin, lay in ruins. But the state was still ambitious. Soon it merged with its neighbour, Prussia, to become a leading power of Protestant Europe. Prussian monarchs turned to Jewish financiers, and new ventures became a source of tax.[9] Despite such success, Jews remained second-class citizens, guest-workers who could only enter Berlin after paying a special levy each time. One who did so in 1743 was a 14-year-old humpback named Moses. His father, a scribe of hand-written Biblical scrolls, was called Mendel, so the boy awarded himself the family name Mendelssohn.

Under Prussia's figurehead of Frederick the Great, the unlikely figure of Moses Mendelssohn (1729–86) went on to become a star of a new Enlightenment, where freedom of religion was at least tolerated, if not encouraged. It became stylish to adopt foreign culture: Voltaire was invited by Frederick with the prospect of a magnificent salary. In this dynamic new intellectual atmosphere, Mendelssohn's early work emphasized the importance of Germany's own legacy, notably that of the polymath Gottfried Wilhelm von Leibniz (1646–1716). Mendelssohn soon became the luminary of the new Prussian intellectual scene, a fully assimilated Jew. But this seems to have pricked his conscience, and he returned to his Jewish roots. Nevertheless, the achievements of the unsightly figure provided a new role-model for upwardly mobile German Jews. One was his own grandson, the composer Felix Mendelssohn Bartholdy (1809–47), whose contemporary was the composer and cellist Jacques Offenbach (1819–1880), the son of a synagogue cantor. But success brought envy and resentment. Soon after Felix Mendelssohn's premature death, the German composer Richard Wagner, writing anonymously, published *Das Judenthum in der Musik (Judaism in Music)* a vitriolic attack on Jews in general and their music in particular.

New liberal trends had been stimulated by the 'liberty, equality, and fraternity' of revolutionary France. Its 1789 Declaration of Rights advocated freedom of religion and the right to worship, sundering the traditional association between rule and imposed religion. Another wave of assimilation began, spreading across Europe in the wake of Napoleon's conquests, and epitomized by the emergence of the Rothschild financier clan. Their 1818 Prussian loan marked the start of a new international bond market. The dynasty soon spun a web of ventures across Europe, developing new forms of financial paper to facilitate the transfer of funds. Such finance could also operate through new methods of communication, first the railways, and then the telegraph.

As these networks evolved, Paul Julius Reuter (1816–99, the founder of the international news agency) discovered a local communications gap between Berlin and Paris which he overcame by using carrier pigeons, transmitting information faster than mail carried by train. Born Israel Bere Josafat, the son of a rabbi in Kassel, Germany, Reuter formally converted to Christianity at the age of 29. For ambitious Jews, formal conversion to Christianity could be seen as a career booster, a benchmark of assimilation. In Berlin in the first half of the nineteenth century, baptism reached an annual rate of a few per cent of the total Jewish population.[10] Having made the decision, the next question was whether to convert to Protestantism or Catholicism. On choosing the latter, one wit pointed out 'there are too many Jews among the Protestants'.[11]

Baptism also provided a useful opportunity to change an antiquated name into something more modern. One such pragmatic convert was Heinrich Heine (1797–1856), born in Düsseldorf with the traditional Hebrew name of Chaim. After initially attending a Jewish school, he tried unsuccessfully to follow family ventures in commerce and banking, and then to make his own career in law. He cynically described his conversion to Protestantism as 'a ticket of admission to European culture: Berlin is worth the sermon'. To avoid censorship, he moved to Paris in 1831. Among those he met there was Karl Marx (1818–83), who had also been baptized, emulating his father. Both Karl Marx's grandfather and uncle had been rabbis in Trier. Heine went on to become one of Germany's most influential poets, his work frequently reflecting his own background.

Berthold Auerbach (1812–82) was once destined for a career as a rabbi. Trying to weave together his threads of Jewish and German culture, he turned to writing, becoming famous for his portrayals of the people and lifestyle of the Black Forest. Later came Franz Kafka (1883–1924), born in Prague in a German-speaking Jewish family. Although not religious as such, he had a keen awareness of the intensity of Jewish culture. In 'later life' (he died aged 40) he enthusiastically began learning Hebrew.

After the defeat of Napoleon, the Jews clung to their recently won emancipation. This rankled others who were more concerned with rearranging national frontiers. In 1819, Heine saw at first hand the anti-Semitic 'Hep-Hep' riots (an acronym of the Latin *Hierosolyma ist perdita*—Jerusalem is lost). The riots were eventually quelled by a national show of force, but they demonstrated that the new emancipation had not eradicated old prejudices.

With emancipation and assimilation came intermarriage. In Jewish eyes, a child of a Jewish mother is deemed Jewish, irrespective of the religion of the father, so that intermarriage does not necessarily reduce the Jewish population. Rich Jewish girls were seen as admirable matches for cash-strapped German aristocrats, or for handsome but impoverished army officers, a practice derided by Heine.[12] Assimilation took many generations to become complete, as the Nazis would later have to acknowledge. Integration, assimilation, baptism, and intermarriage did not guarantee protection from persecution. Undergoing baptism did not remove perceived traditional Jewish traits of intelligence, guile and subtlety.[13] Even in *The Jewish Contribution to Civilization*, written in 1938 with the clear aim of countering contemporary anti-Semitic propaganda, Cecil Roth likened the Jews to 'water spilled on uneven ground, flowing irresistibly into every cranny, where it is absorbed by thirsty soil and ceases to have an independent existence'.[14]

The religious emancipation of revolutionary France was re-enacted on 3 July 1869 when Wilhelm, 'by the Grace of God, King of Prussia, etc' issued an official decree. 'All remaining restrictions on civil and political rights due to the diversity of religious beliefs are hereby repealed. In particular, the ability to take part in national and regional representation and the tenure of public office should be independent of religion'. It was the start of a new era: grand synagogues were built, and by 1875, five per cent of the members of the *Reichstag* parliament were Jews. With Berlin beckoning, and as power struggles between emerging European powers began to resolve themselves, villagers and subsistence farmers left the land in search of new fortunes. The Jewish populations of other capitals such as Warsaw, Prague, Vienna, and Budapest also increased dramatically.

These new cosmopolitan Jews were a new *bourgeoisie*, with access to higher education and new professions, and a vastly improved lifestyle. By the beginning of the twentieth century, a quarter of Prussia's 200 or so millionaires were Jews, although Jews as a whole made up only about one per cent of the total population. The majority of these *nouveau riche* Jews had made their money in finance and banking. In Berlin, 18 per cent of doctors and eight per cent of lawyers were Jewish.[15] But whatever their accomplishments, they seemed to have a driving need to consolidate what still sometimes felt a precarious perch. 'The need to excel was instilled by tradition and nurtured by hostility', wrote the historian Fritz Stern.[16]

However such Jewish success could be resented by others. The new science of heredity and genetics was an opportunity to display any restricted society, like that of the Jews, in a new light.[17] Echoing the sentiments of Wagner, an influential 1879 book by Wilhelm Marr, *Der Sieg des Judenthums über das Germanenthum (The Victory of the Jews over Germany)* seeded the idea of a Jewish conspiracy. In the late nineteenth century, organized anti-Semitism gathered momentum, so that by 1893 a group of sixteen anti-Semitic members formed a visible faction in the Reichstag. The effort stalled, but it had set a precedent.

While accultured and ambitious Jews of Germany had been successful,[18] their brethren further east were not. In the aftermath of

Napoleon's defeat, European territories everywhere were re-distributed. The Russian Tsars acquired a great mass of rural Jewry, largely unsophisticated and unskilled. They were herded into a Pale of Settlement, stretching from the Baltic to the Black Sea. Residence and travel 'beyond the pale' became difficult. Beaten down and humiliated, many Russian Jews became destitute, and had to scrape an income in whatever marginal niche they could find: distilling, pimping, and match-selling were traditional. This made them even more conspicuous and objectionable. Thousands of ragged *Ostjuden* (Eastern Jews) made their way westwards, to seek succour among their richer German counterparts. Scraping a living as hawkers or dealers of second-hand goods, and speaking bad German, they became an embarrassment to their accultured co-religionists, and were later to provide an easy target.

The sorry mass of Russian Jewry left inside its huge ghetto was crushed under a relentless series of callous restrictions and prohibitions. They were prohibited from working elsewhere, and banned from selling alcohol, shutting off another source of income. Young men were pressed into long-term service in the most unappealing parts of the Tsarist Army. Attendance at schools and colleges was limited by a strict quota system. Turning in on themselves, the Jews of the Pale created their own schools in their tiny *shtetlech* (townlets). They were forced to pay bribes for whimsical new documents: travel permits, warrants to allow them to work, or for their children to attend schools. Police would pound on doors during the night and demand to see papers which the people had not yet had time to obtain. Such nocturnal raids evolved into isolated pogroms (from the Russian 'to destroy'), which escalated in 1881 after the assassination of Tsar Alexander II. With the Jews alleged to have been responsible, legislative oppression became direct action: Jews were attacked and murdered. The cruel May Laws of 1882 pushed Jews out of peasant villages into already overcrowded ghetto towns, exacerbating problems due to already limited access to employment and fixed school attendance quotas.

The result was a panic flight out of the country. Shut off from further eastward migration by the Ural mountains and the forbidding expanse of Asian Siberia, these impoverished refugees plodded to Baltic ports, or into Germany. Seeing the lack of success there of their *Ostjuden* predecessors, they continued trudging. Some sought shelter in France. Others went to the North Sea coast and embarked for Britain and its overseas dominions, but the vast majority looked further afield. Around the turn of the twentieth century, more than two million Russian Jews crossed the Atlantic to seek refuge in the United States, the furthest point of the westward Jewish migration which had begun in Roman times. The impact they made there is another story.

In the 1930s, the Nazis had taken note of how successful such harassment could be, and assiduously applied it to the half a million or so Jews under their direct control. But there were millions of Jews in Eastern Europe when Nazi armies invaded after 1939.

NOTES

1. Rutherford, with W.H. Bragg (who had shared the Nobel Physics Prize with his son in 1915), went on to develop a technique which was subsequently patented: Campbell, p.370.
2. Wilson, D. *Rutherford, Simple Genius*, Hodder and Stoughton, London, 1983, p.405.
3. Bryson, Chapter 1. However, his arithmetic is wrong: 500,000,000,000 seconds is 'only' about twenty thousand years.
4. Friedländer, S. *Nazi Germany and the Jews,* Wiedenfeld and Nicholson, London, 1997, subsequently reissued as two volumes, the first being *The Years of Persecution, Nazi Germany and the Jews 1933-39,* Phoenix, London, 2003.
5. American Jewish Yearbook, 2002 http://www.policyarchive.org/handle/10207/bitstreams/17111.pdf.
6. Recounted by Levi in *If This is a Man,* first published in 1947, but the book did not become popular until a second edition was issued some eleven years later. In the USA; the book's title is *Survival in Auschwitz.*
7. His book *Children of the Ghetto* (p.11), Israel Zangwill talks of 'the long dark night of Jewry which coincides with the Christian era'.
8. Sebag Montefiore, S. *Jerusalem, The Biography*, Wiedenfeld and Nicholson, London, 2010.
9. Elon, p.17.
10. Hertz, D. *How Jews Became Germans*, Appendix, Yale University Press, 2007.
11. Elon, p.230.
12. Elon, p.86.
13. Lindemann, p.75.
14. Roth JCC, p.40.
15. Brustein, pp.181, 209.
16. Stern, F. *Dreams and Delusions, the Drama of German History*, Yale University Press, 1999, p.109.

17. Brustein, p.133.
18. This sketch of the recent history of German Jewry uses just a few illustrative examples. For a complete view extending over several centuries, see the four-volume anthology *German-Jewish History in Modern Times*, Meyer, M.A. and Brenner, N. (eds), Columbia University Press, New York, 1998.

2

The rise of German science

As midnight approached over the western Pacific Ocean on 31 December 1999, the world watched as the last hours of the twentieth century slipped away. Poised on the brink of a new millennium, the relentless advance of time seemed to pause in the face of an event horizon which masked the view forward. The unfamiliarity of this new dateline rekindled suspicion and superstition. The century coming to its close had brought nuclear weapons and other agents of mass destruction. It had polluted the atmosphere and sowed seeds for climate change. If that could happen in just one hundred years, what would a whole new millennium bring?

Among its wars, pollution, and other curses, the twentieth century had also brought computers. Finance, transport, communications, electricity distribution, ... the heartbeat of civilization was now governed by machines. Allegedly buried inside them were programs whose mathematics only understood years denoted by two digits. Throwing year '00' into these calculations would give nonsensical answers, said some. Collective hysteria grew that the planet was poised for some major disaster, the 'Y2K' threat, or the 'millennium bug'. Newspaper headlines across the world threatened 'meltdown', 'computer time bomb', or 'ticket to disaster'.[1] Vast resources were marshalled. Special units stood ready. But when 2000 came, a few alarm bells went off in Japan, some ticket machines in Australia refused to work, and several automatic displays showed 19100. Otherwise, the shift into the new millennium was practically seamless. The collective sigh of relief was almost a roar.

A hundred years before, as the nineteenth century had edged into the twentieth, there had been less apprehension: civilization appeared instead to welcome the challenge of a fresh era. Instead of hiding under the skirt of history, the world seemed to rush out to greet the future. A dynamic new modernism thrust aside Victorian dreariness. Paris was its artistic centre, drawing artists from all over Europe: Toulouse-Lautrec,

Picasso, Matisse, Utrillo, Braque ... In music, Stravinsky, Bartok, Debussy, Ravel, and Schönberg were exploring new frontiers. In literature, Anton Chekhov, Thomas Mann, Henry James, and Marcel Proust figured among the pioneers of the new era. As he neared his fiftieth birthday, Sigmund Freud had turned from a successful career as a physiologist to pioneer the field of psychoanalysis. His milestone book *The Interpretation of Dreams* appropriately appeared in 1900. Its ideas flung back heavy curtains of convention, and went on to suggest fresh directions in literature and art.

Initially, these innovations had little impact on everyday life. Most visible at the turn of that century were changes in lifestyle. In streets newly lit by electricity, horse power was being replaced by noisy internal combustion engines. Railway locomotives ran on silent electricity instead of hissing steam. Soon, new machines would even fly through the air. The telephone could transmit speech. Moving images could be admired on cinema screens. Music and voice could be recorded and listened to at leisure. Communication had been liberated from the constraints of physical presence. Beams of X-rays could peer inside the human body. Other beams would soon send communications through 'the ether' without even the need for telegraph cables. At the end of the nineteenth century, even rich people had lived comfortably without interior plumbing, electricity, or automotive power. But in the space of just a few years, this changed. Electricity and running water quickly changed from luxuries to perceived necessities. Underlying all these innovations and driving them forward was a new momentum—science.

Interplay between science and technology was not new. In the mid-eighteenth century, island Britain had been sheltered from the political tumult that shook mainland Europe. New manufacturing industries transformed that country from a rural to an urban economy in an Industrial Revolution driven by steam power. To fuel its factories and build its machinery came new industries—mining, mineral extraction and metallurgy, technologies underpinned by the burgeoning science of chemistry. In the nineteenth-century European quest for national identity, chemistry became a sovereignty issue. Everywhere, scientists raced to discover new chemical elements and give them national labels—germanium, gallium (France), holmium (Stockholm), lutetium (Paris), rhenium (Rhine), scandium ...

The continual advance of science brought a second Industrial Revolution, this time based on electromagnetism. Laboratory curiosities such as dynamos, transformers, and motors mushroomed into the new industry of electrical engineering which turned the wheels of industry, transport, and commerce. Such progress seemed to be enough, and students were even discouraged from turning to science. A few Jeremiahs openly declared that science was a closed book. William Thomson—Lord Kelvin—the giant of Victorian science who had helped pioneer the technique of cable telegraphy, proclaimed 'the future truths of physical science are to be looked for in the sixth place of decimals'.[2] In Chicago in 1894, Albert Michelson, whose measurements of the speed of light would later bring him a Nobel Prize and set Albert Einstein on course to his theory of relativity, confidently declared that in physics 'the grand underlying principles have been firmly established and ... further advances are to be sought chiefly in the rigorous application of these principles'.[3]

In spite of such complacency, science suddenly changed gear after 1900. Several times. In 1905, Ernest Rutherford, then working at McGill University in Montreal, and soon to win the Nobel Prize for Chemistry, gave an invited lecture at Yale University. 'The past decade has been a most fruitful period in physical science,' he said, 'and discoveries of the most striking interest and importance have followed one another in rapid succession. ... The march of discovery has been so rapid that it has been difficult even for those directly engaged ... to grasp at once the full significance ... The rapidity of this advance has seldom, if ever, been equalled in the history of science'.[4] But even the enthusiastic Rutherford could not anticipate what was to come over the next half-century. Measured by the scale of this hindsight, by 1905 relatively little had happened at all!

In this scientific march forward, Germany had chosen its own path. In 1789 Baron Friedrich Alexander von Humboldt (1769–1859), son of the Prussian Royal Chamberlain, inherited enough money to travel to South America, there to make precision surveys and draw accurate charts. His work transformed climatology from folklore into quantitative science. As his heroic efforts brought him acclaim in his native Prussia, his brother Wilhelm (1767–1835) became Prussia's Minister of Education, preparing the way for a new system of schools and higher education. Berlin, the capital of Prussia, appeared to have lost its way after being conquered by Napoleon, and its prescient administrators saw science as pointing the way forward.

Born in Berlin, Hermann Helmholtz (1821–94) had aspirations of becoming a physicist, but was diverted instead into medicine, where training seemed more accessible. During an eight-year spell as an army doctor, he used his spare time to work on his own scientific ideas, as Einstein was to do 60 years later. Using the physiological responses of human eyes, ears, and nerves as a precision laboratory, Helmholtz' studies revealed the principle of the conservation of energy: while energy can be transformed from one form into another—a common example being the conversion of mechanical effort into frictional heat—it cannot be destroyed. After leaving the army, Helmholtz moved through a series of physiology professorships, continually stressing the importance of precision measurement. In 1871 he became Professor of Physics in Berlin. Turning from physiology to inorganic science, he set up a new school of electrodynamics. Heinrich Hertz (1857–94), once a student of Helmholtz, showed in 1885 that high-voltage sparks generated invisible electromagnetic signals—radio waves—which crossed empty space and could be picked up in a remote receiver.

In the same year that Helmholtz moved to Berlin, Prussian Minister-President Otto von Bismarck dissolved the short-lived North German Confederation and declared the Second Reich (Empire) as an heir to the Holy Roman Empire. His armies rapidly acquired new territories and a powerful emergent nation was emboldened by revived images of the feats of Barbarossa and Frederick the Great, and inspired by the music of Beethoven. The Empire became the world's leading scientific power, with Berlin the world's most rapidly growing city: in the nineteenth century its population swelled from 500,000 to four million.[5]

Like Thomas Edison in the United States, Werner Siemens (1816–92) was an entrepreneur as well as an inventor. Leaving school early, he learnt engineering in the Army, and went on to found the industrial empire which brought electricity to a burgeoning nation: motors, dynamos, street lighting, locomotives, and trams. In 1884 he was one of the founders and patrons of the new *Physikalisch Technische Reichsanstalt* (Imperial Physics and Technology Institute). One of its objectives was to consolidate the science of precision measurement, and its first Director was Helmholtz, whose daughter had meanwhile married Siemens' eldest son.

When Helmholtz died in 1894, the directorship of the *Reichsanstalt* was first offered to Wilhelm Röntgen (1845–1923), Professor of Physics at Würzburg. Despite approaching 50, Röntgen turned down the prestigious offer, preferring to continue teaching, together with his special way of doing research, periodically locking himself in his laboratory until he had discovered something. His colleagues called him 'the unapproachable'. Like many of them, he was mystified by the phenomenon of 'cathode rays', a term invented in 1876 by the Jewish scientist Eugen Goldstein (1850–1930). In November 1895 Röntgen locked himself away once more, this time to study the light given off by cathode-ray tubes. In a desperate effort to understand what produced this eerie glow, Röntgen covered his laboratory apparatus with cardboard. In one of the most famous 'eureka moments' in modern science, a fluorescent screen across the room lit up. Whatever was being produced in the tube could also pass through cardboard, and Röntgen soon found it could pass through other objects too. He had discovered X-rays (still called Röntgen rays in German), and early in 1896, newspapers carried photographs showing the bones in Frau Röntgen's hand. Within a few months, doctors all over the world were using the new technique to examine broken bones. Röntgen's rays were show-stopping science, but they only set the scene for the extravaganza to come.

Incandescent lighting, when electricity is driven through a resistive filament, had been discovered in 1878. Electric lighting was a growth industry. To push it forward, scientists needed to understand better how hot objects radiated their energy, and how the wavelength of the emitted light changed with temperature. The eye sees different wavelengths as different colours: when a body is heated, it begins to glow, at first dull red, then 'red hot', gradually becoming a brighter yellow before shining 'white hot'. As well as being careful in his own work, Helmholtz paid attention to what others were doing. As he took on new responsibilities at the *Reichsanstalt*, he ensured that a distinguished professorial chair in Berlin was inherited by Max Planck (1858–1947), a starchy figure whose high collar and black tie still marked him as a product of the nineteenth century when he became famous in the twentieth. Planck too was diligent and painstaking. Like Röntgen, he was well beyond the age at which scientists normally make shattering discoveries: in 1900 he was 42, but his dedication and attention to detail were nevertheless to unlock a main scientific portal to the twentieth century.

The quantum era

In the 1890s, *Reichsanstalt* scientists had continually improved their measurements of radiated energy. Planck stood back from the discoveries and tried to make sense of it all.[6] Others had written down equations which reproduced the laboratory readings, formulae describing *how* energy was radiated, but not *why*. In 1900, Planck tried a different approach. He likened the source of the radiated energy to a swarm of invisible tiny oscillators which vibrated faster as they heated up. Their energy, said Planck's new equation, always emerged as multiples of some fixed amount. (In much the same way, prices have to be rounded to the nearest cent or penny for payment with coins.)[7] These coins of radiation were soon given the tradename 'quantum'.[8] After a hundred years, the word 'quantum' has permeated into the collective consciousness, but still retains an edge of mystery, good enough to use in the title of a James Bond film.[9] But for much of the twentieth century, quantum vocabulary was restricted to textbooks and research papers.

As physicists struggled to understand Planck's quanta, other branches of science were producing new results. Chemists' attention turned from measurements to mechanisms, and German scientists were in the forefront of the new science of thermodynamics.[10] Walther Nernst (1864–1941) went on to earn the 1920 Nobel Prize for his work on thermochemistry. In parallel with his scientific work, the resourceful Nernst developed an incandescent electric light which used a ceramic element. This was soon superseded by the tungsten filament version, but not after Nernst had made a fortune by selling the rights to his invention. In 1904 he became Professor of Physical Chemistry in Berlin, where he soon had the ear of Kaiser Wilhelm II. Nernst impressed on the Kaiser the importance of science for the nation, and fresh scientific institutes grew up in and around Berlin to complement those of the Helmholtz era.

In 1911 the proudly named Kaiser Wilhelm Society (*Gesellschaft*) and its various Kaiser Wilhelm Institutes (KWI) became a new badge of German scientific sovereignty. Outside the university system, but complementing it, these new institutes aimed to give researchers the best possible working conditions, freeing them from formal responsibility for teaching. At first aimed at the pure sciences—Biochemistry, Biology, Biophysics, Medicine, Physical Chemistry, and Physics, others for applied research followed—fibres, coal, leather, silicates, and stockbreeding, among many others. Two of the three founding institutes—Physical Chemistry and Medical Research—had Jewish directors, Fritz Haber and Richard Willstätter respectively. There would soon be a third: Albert Einstein at a new Physics Institute.

Meanwhile Bismarck's Second Reich, envious of the vast colonial empires of Britain and France, had established footholds in Africa and Asia. But with unclaimed areas of the globe increasingly difficult to find, Germany sought a fresh route to imperial power. New techniques and materials introduced through the nation's scientific prowess made German industry less dependent on raw materials imported from the colonies of its neighbours. German science became the key to boosting the nation's reputation, its scientists the collective Columbus, setting out to discover a new world and create a technological empire. The KWI's mission statement evoked 'the true land of boundless opportunities'.[11] This effort further consolidated Germany's position as the world's leading scientific nation. In the fifty years prior to the First World War, some ten thousand US students were educated in German universities.[12] German had become the default language for science, and aspirant scientists in other nations were advised to learn it.

The German word for science is *Wissenschaft*, freely translated as 'knowledgedom'. Underlying all this effort of innovation and discovery, and the importance of precision measurements, was mathematics, the ultimate instrument of science. A pioneer had been Gottfried Leibniz (1646–1716), who had discovered infinitesimal calculus at the same time as Isaac Newton in England. The focus of German mathematics became the great university of Göttingen, founded in 1717 by Duke Georg-August of Brunswick-Lüneburg and Elector of the Holy Roman Empire (known elsewhere as King George II of Great Britain). The father-figure of Göttingen's and German mathematics was Carl Friedrich Gauss (1777–1855), known as 'The Prince of Mathematicians', who himself called mathematics 'the queen of sciences'.

Gauss made physics more mathematical, and mathematics more physical. Professor of Physics alongside Gauss at Göttingen was Wilhelm Weber (1804–91), and the pair initiated a new synergy. With Weber, Gauss turned his mathematical attention towards the emerging science of electricity and magnetism. Thus the name of Gauss is now as familiar to physicists as it is to mathematicians: the gauss is the fundamental unit of magnetic field. Daniel Kehlmann's successful novel *Measuring the World (Die Vermessung der Welt)*[13] imagines the lively collaboration between

Gauss and Humboldt. Gauss' successor at Göttingen was Johann Dirichlet (1805–59), who married Rebecca Mendelssohn Bartholdy, a sister of the composer Felix Mendelssohn. After Dirichlet, Göttingen's dynasty continued with the genius of Bernhard Riemann (1826–66), the implications of whose work would extend far beyond the realm of abstract mathematics, and go on to guide thinking in the twentieth century.

Since the time of the Ancient Greeks, geometry—the mathematics of space—used a toolkit of straight lines of various lengths to construct triangles, rectangles, and other shapes. After René Descartes in France and then Gauss had looked for new approaches, Riemann discarded the old Greek rigidity and recast this geometry in terms of calculus, providing a naturally curved framework that later would become Einstein's workspace. After Riemann died of tuberculosis at the age of 39, geometrical science at Göttingen continued its advance with Felix Klein (1845–1929). Klein and his Prussian patrons realized the industrial importance of mathematical science, and pushed for new institutes for applied research in areas such as engineering and metrology.

In 1895, Göttingen's crown passed to David Hilbert (1862–1943), who for many was the greatest mathematician of all. At the International Mathematical Congress in Paris in August 1900, he famously listed 23 intransigent problems which needed to be solved in the coming century.[14] As well as a challenge to the world, the list was also a challenge for Göttingen, and above all for Hilbert himself. One of his challenges advocated putting the whole of physics on a firm mathematical footing. Hilbert thus became a mathematical physicist, using Nature as the natural stage for his mathematics. In 1902 he was invited to become professor at Berlin.[15] Playing his cards skilfully, Hilbert said that he would condescend to stay in Göttingen only if an additional chair of mathematics were established. For this, he recruited Hermann Minkowski (1864–1909).

Minkowski was born in 1864 into a family of German Jews living in what is now Lithuania, but who soon moved to Königsberg in Prussia. It was a talented family: Hermann's brother Oskar went on to discover that diabetes was a malfunction of the pancreas. But Hermann died of appendicitis before his mathematical collaboration with Hilbert had yielded all of its fruits. Into the scientific vacuum left by Minkowski's untimely death would soon step an unexpected newcomer, Albert Einstein.

Minkowski was not the first distinguished German Jewish mathematician. Carl Jacobi (1804–51) initially qualified as a schoolteacher, but continued with higher mathematics, travelling across Europe and meeting with great names—Legendre, Fourier, and Poisson. He eventually became Professor at Königsberg, and his name is commemorated through the Jacobian matrix, a standard mathematical tool. Other early Jewish German mathematicians included Georg Cantor (1845–1918), the creator of modern set theory, and the author of the 'Continuum Hypothesis' invoked by Hilbert as a mathematical challenge, and Leopold Kronecker (1823–91), who made a lifetime study of the importance of integers: 'God made the integers, all else is the work of man'.[16]

Chemical force

In the nineteenth century, Germany also acquired chemical sovereignty. The combined strength of new chemical concerns—Bayer, founded in 1863, Hoechst (also in 1863), BASF (*Badische Anilin- und Soda-Fabrik*, 1865), and Agfa, (*Aktiengesellschaft für Anilinfabrikation* 1867)—went on to lead the world. This predominance was consolidated when German chemists developed powerful new manufacturing methods. Adolf von Baeyer (1835–1917) was awarded the 1905 Nobel Chemistry Prize for his synthesis of the dye indigo, but the overriding contribution came from Fritz Haber (1868–1934). Born in an assimilated Jewish family in Breslau (Wroclaw), Haber did everything possible to foster his acceptance as a German. At the late age of 24, he was formally baptized. At first, this move met with little academic success, which one of his colleagues attributed to the fact that at under 35, he was too young for a university post, at over 45 he was too old, and between 35 and 45 he was a Jew.[17] His fortune changed in 1907 when he developed what became the standard industrial method for synthesizing ammonia from atmospheric nitrogen (the Haber process). This made Germany independent of imported materials for agricultural fertilizers, and, on the eve of war, for explosives too. This success, underlined by the 1918 Chemistry Nobel Prize, made him a national figure.

A wave of patriotism and industrial and commercial success had helped fund the new Kaiser Wilhelm Institutes. A prominent Jewish benefactor of the Institute for Physical Chemistry made his contribution conditional on Haber being made its Director. But when war broke out in 1914, Haber, despite being 46, joined the Army. Regardless of his enthusiasm and his elevated position in civil life, he was initially only a lowly sergeant in the German Pioneer Corps: the emancipation of Jews had not yet spread to the military. This was rectified for Haber when the German army needed chemists to develop a new weapon of mass destruction—poison gas. The first gas to be used was chlorine, much heavier than air, carpeting the ground, and spilling into trenches. In April 1915, on a four-mile stretch of the Ypres salient, soldiers who stayed in their dugouts suffocated horribly, while others who fled became open targets.

In these first chlorine attacks, a young German lieutenant named Otto Hahn supervised the installation of gas cylinders. Like Haber, Hahn went on to win a Nobel Chemistry Prize. Other contributors to the wartime German gas effort who later became Nobel laureates were Walther Nernst, as well as James Franck and Gustav Hertz, of whom more later. Together, it was an impressive array of scientific expertise for such an evil purpose. In a paroxysm of diabolical innovation, German chemists in industry and in the Kaiser Wilhelm Institute sought new gases, like phosgene, fatal with much lower concentrations than chlorine, and which could be compressed into artillery shells. New gas cocktails included a vomiting agent, making soldiers rip off their gasmasks and exposing them to other deadly agents. Then came mustard gas, which attacked slowly, causing horrible blisters on skin and eyes, blinding, and with severe trauma if the gas was inhaled. To disguise their lethal cocktails, ingenious chemists added perfumes which smelled like flowers.[18]

In 1901, Fritz Haber had married Clara Immerwahr, the first woman to earn a doctorate at the University of Breslau. Like her husband, she had formally abandoned Judaism. Ignored and sidelined by him, and dismayed by his devotion to such evil work, she demanded that he stop. When he refused and abruptly left for another gas warfare installation, she shot herself with his pistol.[19] In 1918, the victorious Allies, despite themselves having adopted the same gas tactics, wanted to try Haber as a war criminal. However their efforts were blocked by his exalted international status following his 1918 Nobel Chemistry Prize, awarded for his industrial synthesis of ammonia, conveniently overlooking his involvement with poison gas. Depressed and dysfunctional despite his Nobel award, widower Haber later devised a scheme to attempt to extract gold from seawater and boost the ruined finances of post-war Germany. Unlike ammonia synthesis and poison gas, it did not work.

While Haber had been discovering how to manufacture ammonia, in Manchester another chemical synthesis had important implications for the coming war effort. Chaim Weizmannn was born in 1874 into a family of religious Jews in the Tsarist Pale. Moving to Germany to study, he was attracted by the new Zionist movement to establish a Jewish homeland in Palestine, but initially continued his scientific career. As a university biochemist in Manchester, he focused on the problem of finding a source of synthetic rubber, another vital industrial commodity dependent on a long and precarious chain of imports. His progress illustrates the unpredictable serendipity of scientific research: for synthetic rubber, he first needed synthetic butanol. In developing a bacterium to ferment butanol from corn, Weizmann was led to produce acetone, a valuable raw material for wartime high explosives. During the war, Weizmann's attention was shifted by the 1917 Balfour Declaration—the British government's backing for Zionism in Palestine (soon to come under British control). Dedicating himself to the new cause, Weizmann led the international Zionist Movement, and became the first President of the new state of Israel when it was established in 1948.

In parallel with such industrialization of chemistry, medicine saw the emergence of new techniques such as bacteriology and immunology. Paul Ehrlich (1854–1915) was born into a Jewish family with a tradition of distilling, but soon became attracted to science. In his research career, he helped develop antitoxins to combat dangerous diseases, in particular cholera, diphtheria, and tuberculosis, and went on to earn the 1908 Nobel Prize for Medicine. Later, he developed a new drug, Salvarsan, which became the major weapon against syphilis until the development of antibiotics several decades later. Ehrlich also pioneered the technique of chemotherapy, using 'magic bullet' drugs. Ehrlich retained his allegiance to Judaism and was never baptized to smooth his way into conventional academia.

Ehrlich's daughter went on to marry the distinguished German-Jewish mathematician and number theorist Edmund Landau (1877–1938), who after teaching at the universities of Berlin and Göttingen, emigrated in 1925 to Palestine to establish a mathematics school at the Hebrew University of Jerusalem. Although a gifted mathematician, he was politically outmanoeuvred in his efforts to get the new department off the ground. Soon after he returned to Göttingen, the new Nazi laws banned Jews from university posts, and Landau left Göttingen in 1933; a premature retirement.

Stigma and enigma

To help build a mathematical foundation for the axioms of physics, Hilbert at Göttingen had enlisted the help of Emmy Noether, born in Erlangen in 1882, the daughter of an accomplished Jewish mathematician. After secondary school, Emmy Noether prepared for teaching,[20] then the standard career for intellectual women, but under her father's influence continued with mathematics at Erlangen, where she was one of two women students on a campus with a thousand men. In the early years of the First World War, the universities suddenly emptied. Women would not have to leave for the front, and Hilbert invited Emmy Noether to Göttingen. A vociferous liberal, Hilbert pushed for the university's admission rules to be made more flexible. Before teaching at a German university, candidates first have to pass a 'habilitation' exam,[21] and the Prussian authorities had explicitly excluded women, even as candidates. Hilbert objected. While full equality of opportunity was not yet on the agenda, exceptions should nevertheless be made, he demanded.

A century ago, women academics were treated by their male counterparts with a formal contempt which is now inconceivable. Thus when Emmy Noether's case was brought to the full university court, tempers rose and Hilbert made the famous remark that Göttingen was 'a university, not a bathing establishment'.[22] His plea was refused. Undaunted, Hilbert went to Berlin to discuss the case with the Ministry and negotiated a concession under which Emmy Noether could 'support' his teaching. In practice this ran to actually giving courses. In 1919 Emmy Noether became a private tutor (*privatdozent*), and in 1922 achieved the lofty status of non-official professor with no salary and no tenure. Her income still came from private tuition. Despite job offers from Moscow and Frankfurt, she preferred the dynamism of Göttingen.

Unusually for a mathematician, Emmy Noether's most important work came fairly late, as she was nearing 40. To tighten Einstein's relaxed mathematics, Hilbert maintained that the formalism of relativity had to be more rigorous. Emmy Noether was one of the few who could handle these techniques. In 1918, as the argument about her formal status at Göttingen still raged, she showed how to resolve the dilemma. All conservation laws, such as that of energy pointed out by Helmholtz, could be elegantly embodied in a succinct mathematical theorem. Einstein was immediately enthusiastic: 'I have studied your paper ... with true amazement. ... Everything is wonderfully transparent'.[23] Despite these successes, her job continued to be precarious. After a visit to Russia in the winter of 1928–29, as if being a woman academic was not enough of a burden, on her return to Germany she was branded as a Bolshevik as well as a Jewess. Her growing band of supporters hoped that awards and celebrations of her fiftieth birthday In 1932 might finally mark some kind of watershed of acceptance.

It was not to be. In 1933, new Nazi laws put hard-won Jewish emancipation on fast rewind. Emmy Noether lost even the marginal acceptance she had been grudgingly accorded, and moved to Bryn Mawr women's college, Pennsylvania, with money from the Rockefeller Foundation. She returned to Germany briefly to see her brother Fritz, also a mathematician, before he emigrated to Tomsk in Siberia. Soon after her return to the USA, in 1935, Emmy Noether died from complications following what would now be considered a routine operation for the removal of a tumour. To compound the family misfortunes, in 1941, her brother Fritz was arrested in Tomsk, charged with espionage, and sentenced to 25 years' imprisonment. He was shot in 1941. Although recognized by Einstein and some other mathematical scientists, the work of Emmy Noether was unknown to her physics contemporaries. Only later did its implications become clear: 'Noether's theorems' became a touchstone of physics, and remain as a monument to a tragic figure, an intellect ahead of her time and a multiple victim of prejudice.

When she was forced to emigrate, Emmy Noether had had two options. Either she could go to Somerville College, Oxford, or to Bryn Mawr. As well as offering more money, the American option was also a sign of intellectual climate change. European academia had its traditional stepping stones: at the top were the prestigious universities: Göttingen, Berlin, Oxford, Cambridge, Paris. ... There, competition was intense, both for student places and for teaching jobs. Across the Atlantic, the United States had its own premier Ivy League universities, but its nineteenth-century scientific education still acknowledged the supremacy of Europe. Nevertheless a few tiny pointers had appeared.

At Yale, Josiah Willard Gibbs (1839–1903), described by Bill Bryson as 'the most brilliant person most people have never heard of',[24] helped build the new science of thermodynamics. Another who caused US science to be noticed was Albert Michelson (1852–1931), born into a Jewish family in Strzelo, in Prussian-occupied Poland, but who came to the United States with his parents before the panic flight precipitated by the pogroms. The family went to California, becoming shopkeepers in remote gold-mining towns.[25]

Seeing that their son had talent, the Michelsons sent him to school in San Francisco, where he was advised to try a competitive entrance exam for the Naval Academy in Annapolis. He just failed, but reluctant to accept such a narrow decision, travelled across the country to Washington DC. There, he waited outside the White House to intercept President Ulysses S. Grant as he emerged for a stroll.[26] Despite the recent assassination of Abraham Lincoln, security was not considered important in Washington in those days: public figures really were public. Either because Grant realized the importance of the Jewish vote, and/or because he was worn down by Michelson's insistence, he duly accorded his suppliant a place at Annapolis.[27]

After several years' service at sea, Michelson returned to shore to teach at the Naval Academy. While instructing sailors in basic mechanics, he glimpsed a more demanding scientific challenge—to measure the speed of light. It had long been believed that light is transmitted instantaneously, with infinite velocity. But seventeenth-century astronomers tracking the moons of Jupiter had found effects which suggested that light has instead a definite velocity. In 1849, Hippolyte Fizeau in France ingeniously measured the speed of light for the first time without having to look up at the night sky. Light travels about 300 metres per microsecond, and measuring something that moves so fast calls for ingenuity and painstaking attention to detail. However precision measurements were Michelson's idea of 'fun',[28] and setting himself the task of improving on these initial measurements, he first travelled to Europe to learn how scientists there did it. Among other places, he visited Helmholtz' laboratory in Berlin.

In 1887,[29] his milestone measurements showed that the velocity of light always appeared to be the same, independent of the speed of its source. This was a complete mystery. A passenger falling from a moving vehicle hits the ground with the combined downward speed of the fall and the forward speed of the vehicle. Everyone assumed that a similar effect would happen for light, so that light from an object that was moving, for example because of the west–east rotation of the Earth, would not travel at the same speed as light from a more 'stationary' source pointing north–south. Michelson showed otherwise: the velocity of light had no such directional effect.

Although this unexpected result plunged science into a dilemma, it nevertheless brought Michelson scientific fame, and he moved up through a series of university positions before being invited to create a physics group in a new University in Chicago, established under Rockefeller patronage. With no Ivy League background, Michelson remembered his experience in European universities. Chicago physicists went on to play a major role in the development of science in America and in the world. In 1907, Michelson was awarded the Nobel Prize for Physics for his precision work in optics, the first citizen of the USA to receive one, and the first Nobel laureate with full Jewish ancestry.[30]

Michelson's speed-of-light conundrum had to wait eighteen years for an explanation. It came with Albert Einstein's revolutionary theory of relativity. At the beginning of the twentieth century, the constancy of the speed of light was one of two great scientific mysteries: the other was Max Planck's enigmatic new quantum picture. Both were soon explained by Albert Einstein, the apostle of twentieth-century scientific modernism. Einstein could grapple with the incomprehensible and force it to make sense. In the space of a few years, this uncanny ability plucked him from anonymity and turned him into an almost venerated figure. For fourteen years, from 1919 to 1933, he was an icon of science in Germany, a land which itself had already become a focus of world science in an increasingly scientific century. His scientific achievements won him many admirers, but this modernist science was so revolutionary, so baffling, that some scientists rejected such impertinence.

Einstein solved the enigma of the constant velocity of light by turning the problem upside down. The basic premise of his new theory of relativity is that light necessarily travels at the same speed—no information can pass more quickly than a light signal. What had been an enigma became instead a central tenet, a new foundation stone of science. But cementing this in place soon displayed cracks in the existing infrastructure. What were theorems for Einstein became paradoxes for others: one of a pair of twins choosing a career as a high-speed space traveller would age at a different rate to his terrestrial counterpart.

After creating relativity with his new view of the role of light, Einstein's second contribution to relativity was paradoxically less controversial because it was more difficult to understand. In 1907, he had what he later called '*Der glücklichste Gedanke meines Lebens*' ('the most fortunate thought of my life'). He realized that the force of gravity, which keeps our feet on the ground and the planets in their orbits, has only a relative existence: a person falling freely in space is 'weightless'. After absorbing the implications of this idea and learning the mathematics of Bernhard Riemann, Einstein forged a new explanation of gravity as the natural result of the geometry of space. At first, few cared, but in 1919 a solar eclipse allowed astronomers to check Einstein's new prediction of how the light from distant stars is bent by gravity as it grazes past the Sun. 'Lights All Askew in the Heavens', proclaimed the New York Times. Rarely has a scientific theory that so few could understand made such a collective impact.

After Einstein had thought his way through the Universe, its cosy familiarity had been replaced by new incomprehensibility. This enduring paradox was the key to Einstein's fame. He himself revelled in his visibility. This was further cemented by the award of the Nobel Physics Prize for 1921 'for his services to theoretical physics'. The citation said little about relativity, because the authorities in Stockholm who made these decisions were themselves perplexed and unsure. Einstein's Nobel citation focused instead on his resolution of Planck's quantum enigma. When light shines on photosensitive materials, it knocks electrons out of atoms, creating a tiny electric current. To explain this 'photoelectric effect', in 1905 Einstein proposed that light is not a continuous stream on the microscopic scale: instead, it is made up of Planck's 'quanta', individual bullets of radiation. What had been introduced by Planck five years earlier as an astute mathematical ploy now became a concrete mechanism.

The next step was to reconcile this new granular picture of light with its traditional interpretation as continuous electromagnetic waves rippling through space. It took another twenty years or so for the full new duality to emerge through the work of Schrödinger and Heisenberg, and in 1927, scientists found that reflected electrons showed wavelike patterns. At this microscopic level, particles and waves are not intrinsically different, but are two ways of looking at the same thing.

Incomprehensibility was the hallmark of this new 'quantum mechanics'. As Niels Bohr, one of its ardent advocates, enigmatically put it: 'anyone who is not shocked by quantum mechanics has not understood it'.[31] It certainly baffled Einstein, who revelled in his bewilderment. Just as others had tried to highlight apparent loopholes in his relativity, he tried valiantly to demonstrate that quantum mechanics was inherently wrong. In 1935 he pointed to what he thought was a paradox,[32] but which instead went on to open the door to even more bizarre effects: such 'quantum entanglement' now forms a basis for modern quantum computing and cryptography.

But Nobel committee members were not the only ones who doubted Einstein's science. In Germany, strident voices denounced the emergence of all forms of unintelligible modernism and derided the abandonment of traditional values. This included science, and anything that flouted 'commonsense' rules was condemned. Einstein's work was at the top of this list. Although relativity and quantum theory were the harbingers of new twentieth-century science, they were also to hasten the twilight of a century of Germany scientific supremacy.

NOTES

1. Davies, N. *Flat Earth News*, Vintage, London, 2009, p.10.
2. Barrow, J. *Impossibility, The Limits of Science and the Science of Limits*, Vintage, London, 2004, p.54.
3. Grometstein, A. *The Roots of Things, Topics in Quantum Mechanics*, Springer, Berlin, 1999.
4. PaisIB, p.128.
5. Mendelssohn, p.50.
6. Mehra, J. Max Planck and the Law of Blackbody Radiation, in *The Golden Age of Theoretical Physics*, Word Scientific, Singapore, 2001.
7. For major international payments, currency exchange rates are quoted to four places of decimals.
8. Rhodes MAB, p.70. However the word, from the Latin for 'how much?' had been used in other contexts in the nineteenth century.
9. *Quantum of Solace*, released in 2008.
10. Ludwig Boltzmann was Austrian, but worked for four years in Munich.
11. Medawar and Pyke, p.4.
12. Kirschbaum, E. *The Eradication of German Culture in the United States*, 1917–1918, H. D. Heinz, Stuttgart, 1986.
13. Kehlmann, D. *Measuring the World*, Quercus, London, 2007.

14. Devlin, K. *The Millennium Problems*, Basic Books, New York, 2002.
15. Mehra, J. The Göttingen Tradition, in *The Golden Age of Theoretical Physics*, World Scientific, Singapore, 2001.
16. Bell, E.T. *Men and Mathematics*, Simon and Schuster, New York, 1986.
17. Mendelssohn, p.84.
18. Rhodes MAB, p.90 and references therein.
19. Cornwell, p.65.
20. Byers, p.83.
21. *Habilitation* is the highest level a student can attain through study.
22. Byers p.91.
23. Byers, N. in *History of Original Ideas*, ed Newman, H. B. and Ypsilantis, T. Plenum, New York, 1996, p.950.
24. Bryson, p.154.
25. Just a few years earlier, another Jewish emigrant had moved to California. Levi Strauss (1829–1902), born in Bavaria, went to the USA at the age of eighteen and joined the New York fabric business established a few years earlier by his elder brothers. To profit from the gold rush, he went to California to open a new branch. There, he marketed the denim working garments which later became an international fashion item.
26. Bryson, p.156.
27. Across the Atlantic another young Jew was about to embark on a military career. Alfred Dreyfus (1859–1935) became an officer in the French army in 1890, only to fall victim to a vicious anti-Semitic conspiracy in 1893. Alleged to have been a spy for Germany, he was incarcerated on Devil's Island in the South Atlantic. After a public outcry and retrials which traumatized France, he was later pardoned.
28. Pais SL, p.116.
29. with Edward Morley.
30. Adolf von Baeyer, the German 1906 Chemistry Laureate, had a Jewish mother.
31. Gaither, C. *et al.* (eds), *Physically Speaking, A Dictionary of Quotations on Physics and Astronomy*, Taylor and Francis, New York, 2007.
32. The Einstein/Podolsky/Rosen paradox.

3

Cultural cleansing

At the eleventh hour of the eleventh day of the eleventh month of 1918, guns which had been firing for more than four years fell silent. On the Western front, from the Vosges to the sea, men climbed out of their trenches and dugouts, stood up in the open and cheered. No longer would uncomprehending generals bark orders that condemned obedient troops to mass slaughter. Almost ten million had been killed. The scale of war had changed; in the previous European conflict in 1870, some six thousand soldiers had perished at the battle of Sedan. Between 1914 and 1918, about the same number were killed on average every day. In those four years, the world had altered. Maps would have to be redrawn. Four once-mighty empires were gone: the dominion of Tsarist Russia; the Imperial Germany forged by Bismarck; the once-proud Austria-Hungary of the Habsburgs; and the vast but fragile Ottoman Empire of the Moslem Caliphs. With these changes, cultures crumbled and viewpoints were radically shifted.

The events of 1918 were even more remarkable by their suddenness. After Russia had fallen into civil war, German leaders congratulated themselves on a victory in the East. Troops were rushed westwards to reinforce a German offensive which at one time had threatened to break the war's long stalemate. But it came to nothing: the Allied army hit back with forces which now included tanks, air cover, and increasing numbers of fresh American soldiers. German resistance crumbled: battle lines across France and Belgium that had hardly moved for years melted back towards the Rhine. Within Germany, the population already had been demoralized by privation, even starvation. Kaiser Wilhelm scuttled across the border into neutral Holland, leaving embittered generals and confused politicians to squabble over a mess that was now social and economic as well as political. Unrest had become anarchy.

Such an ignominious and incomprehensible defeat was especially bitter for a nation which still remembered its greatness. Bismarck's

Second Reich could not rival the empires of Britain and France in geographical extent, but had not lagged behind its European neighbours scientifically, economically, and industrially.[1] Life in Germany in the years immediately before the war had been memorably good. But the defiant country that had cheered the declaration of war in 1914 had lost millions to gunfire, bombs, malnutrition, and the 1918 influenza epidemic. Two million German soldiers had died on the battlefield, more than any other nation involved in the conflict.[2] Germany was humiliated, hungry, and resentful, but its enormous industrial infrastructure was still largely intact. The Second Reich had evaporated, but what was to take its place? Millions of dejected German soldiers plodded home, still technically a fighting force.[3] The war had ended in an armistice, not a surrender.

Military mutinies sparked anarchy within the borders of a nation which had learned of the war via reports from sons and husbands fighting abroad. In the vacuum left by the Kaiser's hurried departure, Bolshevism, which only one year before had overthrown the Russian Tsarist rule that had lasted for centuries, provided one compelling new political option. The various grievances of the industrial working class had set the scene: German socialists, many of whom had spent the war years in prison, sparred with frustrated nationalists on the right and social democrats in the centre. In the confusion, nationalist militias fought with mutineers, and socialists with aristocrats. The army, ordered to attack mutineers, mutinied again. A general strike added to the national mayhem.

During the war, the British navy had sealed up the North Sea, cutting off Germany's main supply route for precious raw materials and foodstuffs. The attrition of the resultant malnutrition had sapped the energy and undermined the health of the population. Ingenious *ersatz* substitutes had appeared: *Kriegsbrot* (black war-bread) used potato meal and had the advantage of not going stale. Such penury bit particularly hard in the towns and cities, with little direct access to fresh produce. Callously, the British maintained this vice-like blockade until the armistice was formally signed in 1919. In spite of all these problems, the normally perceptive Albert Einstein in Berlin penned an optimistic postcard to his mother: 'militarism and bureaucracy have been thoroughly abolished here', he wrote.[4] It was not one of his better observations.

Amid all this, the role of Jews in German politics became more noticeable, with radical Polish and Russian Jewish immigrants reinforcing the extreme left wing of the socialist movement. During the war, Jews had become more conspicuous. Firstly, the Jewish population had grown. Millions of men under arms made for labour shortages, and with Germany temporarily occupying territory in Russian Poland, about 100,000 Eastern European Jews (*Ostjuden*) moved westwards to try to improve their lot and escape from the continual threat of pogroms.[5] In 1914, the Jews had been as enthusiastic as other Germans when the nation went to war. As the country was gripped by a frenzy of patriotic fervour, Jews such as Fritz Haber had been quick to enlist and offer their services. Some 100,000 in a population of half a million saw active service: 12,000 of them died. Jewish visibility had also increased when they became prominent in the new Social Democratic Party, which came to power in the 1912 *Reichstag* elections: 20 out of a total of 25 Jewish *Reichstag* members represented the socialists.[6]

After several false starts, a republican assembly—a novelty for Germany—convened in the central town of Weimar. But this did not resolve the turmoil. Towards the end of the war, socialists under Kurt Eisner—a Berlin Jew—had replaced the monarchy in Bavaria. Eisner's precarious mandate as premier of Bavaria ('The Jewish Republic')[7] only lasted a few months. However mere resignation was not enough: his subsequent assassination by a right-wing aristocratic extremist made him a martyr. Another minister, assumed to have been in some way responsible, was shot. In 1919, Eisner's emotional funeral seemed to push Bavarian and German politics over a precipice. Rekindled violence led to a Soviet-style Republic being declared in Munich, bringing more political chaos elsewhere. One notable victim was the Jewess and Marxist-Communist figurehead Rosa Luxemburg. Born in the Polish/Russian Pale of Settlement, she became a German left-wing activist. In the First World War, she shouted for non-belligerence and was imprisoned. Released after the war, she fronted renewed agitation in Berlin, where she was arrested and shot without trial.

Violence bred more violence as hotheads sought to find handholds to power from the ruins of defeat. In March 1920 socialist and nationalist mobs confronted each other in Dresden, and a shot fired inside the palatial Zwinger art gallery drilled a hole in a painting by Rubens. The Austrian painter Oskar Kokoschka told citizens that if they had to fight, they should do so where 'civilization is not at risk'.[8] He added, 'pictures cannot run away from places where human protection fails them'.

His words went unheeded, and the protection was not to be confined to inert pictures.

An Austrian infantryman called Adolf Hitler, who had been temporarily blinded by gas in Belgium, had returned to his regiment in Munich. There he saw Eisner's funeral and was drawn into underground politics. In trying to make sense of the chaos of the war, everyone had their favourite reason and their preferred scapegoat. The disillusioned soldiers blamed their generals, the generals blamed the Kaiser, and the monarchists blamed the communists, but the movement to which Hitler was drawn found a new culprit—the Jews. Although they had played their role in the war, dissidents like Rosa Luxemburg apparently became more visible. As the conflict continued to grind on remorselessly despite Haber's poison gas, some Germans began to suspect the futility of such a conflict. The press, largely Jewish-owned, voiced a growing call for a negotiated armistice. Such demands fuelled resentment, and antagonized patriots and the military. After the armistice, and with much of the nation's industrial capability still apparently intact, uncomprehending Germans searched for an explanation for their misfortune. The 'stab-in-the-back' (*Dolchstosslegende*) myth was born: a sinister conspiracy had not only ignored the call to arms but had systematically undermined the war effort and even sabotaged it. Allegations flew that while noble Germans fell on the battlefields, greedy Jews made profits from the war.[9] These accusations gradually transformed from vague supposition into a conviction which spread to the body of the nation. It had happened before in Germany: the Hep-Hep riots had broken out in 1819 in the aftermath of another war. Sensing another such change in sentiment, Jews tried to make their position more secure: the baptism rate for Jewish males had increased from 8.4 per cent in 1901 to 21 per cent in 1918.[10]

After the British Navy finally relaxed its blockade, the nation became hamstrung by mercenary demands from the victorious powers for reparations and material indemnities. Germany was to hand over 5000 locomotives, 150,000 railway wagons, and 5000 trucks, and comparable amounts of military equipment.[11] But even this bounty was overshadowed by outrageous demands for continued cash payments over the next half a century. Faced with such extortion, and having lost swathes of territory and whole armies of manpower, German industry stalled, and the impoverished nation defaulted on its huge repayments. In 1923, French forces moved into the Ruhr. As they seized control of Germany's industrial heartland and workers walked off their jobs, Germany was no longer producing enough goods. With the victors still clamouring for cash, Germany began to print worthless money. This sparked a wave of hyperinflation which is still remembered whenever fragile economies start to get jittery.

In 1918, the US dollar bought 4.2 German marks. By 1923, this exchange rate had ballooned to a trillion. A pound (half a kilo) of bread cost two billion marks, the same amount of meat 36 billion, and a glass of beer four billion. It was easier to heat houses by burning wads of worthless banknotes than going out to buy firewood. Prices would double every few days. One contemporary joke was about a patient discharged from a mental hospital who took a taxi home:

PATIENT: How much is the fare?
TAXI DRIVER: 4000 million marks.
PATIENT: But I don't have that much money!
TAXI DRIVER: How much can you give me?
PATIENT: I only have this old gold ten-mark piece.
TAXI DRIVER: Wonderful, here's your 8 million change.
PATIENT: Keep the change. Take me back to the asylum![12]

By this time, there had been hundreds of political assassinations by extremists of all kinds. From all this political ferment, the German National Socialist Workers (Nazi) Party, now with the 30-year-old war veteran Hitler as its head, slowly emerged as one counterweight to Bolshevism. The Nazis openly brandished their motto of anti-Semitism: the Jews were alleged to be social and financial vampires who had degraded Germany. The financial humiliation of hyper-inflation had fuelled German resentment, especially when they could glimpse foreigners with hard currency enjoying a lifestyle which had become increasingly inaccessible with worthless marks. Rabble-rousing rant accused Jewish financiers. And there was the political angle: for the Nazis, Jew and Bolshevik were practically synonymous.

A traditional German undercurrent of soft anti-Semitism was being whipped into a storm by the radical new Nazi mind-set. The transformation to violent anti-Semitism took a tragically palpable form in 1922 when the newly appointed Jewish Foreign Minister, Walther Rathenau, was machine-gunned in a Berlin street by right-wing gangsters. Born in

Berlin in 1867, Rathenau was the son of the founder of the AEG electrical engineering combine, and during the war had held key positions in the industrial effort which had kept the nation going. Sensing what was coming, he had urged German Jews to strive to be less conspicuous and to assimilate, convinced that only this way would they survive.[13] Germany was shocked by his assassination, and his body was accorded the pomp of a state funeral. A few days after the murder, another prominent Jew, newspaper editor Max Harden, was beaten up. He had advocated that only by strictly adhering to the harsh terms of the post-war armistice agreement could Germany hope to take its new place in the world. In the ensuing trial of his attackers, Harden was deemed a defeatist and was vilified almost as much as his assailants. In 1923, with increased support from a Bavarian establishment, Hitler staged his famous Munich demonstration: the police opened fire on the demonstrators. Arrested and imprisoned, he wrote his iconic book, *Mein Kampf*, and was freed after less than a year.

The aftermath of the war had also left its mark on popular culture. In the 1920s, Berlin threw off its shame as the capital of a defeated nation, becoming instead the focus for a new sensual lifestyle, the world capital of permissiveness. A new wave of modernism emerged, with *Bauhaus* design under Walter Gropius, Dada art with Max Ernst, and atonal music with Arnold Schönberg. A major success was Bertolt Brecht's *Threepenny Opera (Dreigröschenoper)*, whose music by Jewish composer Kurt Weill included the now classic song 'Mack the Knife'. German cinema boomed, with Marlene Dietrich as its cult figure. In the public imagination, the fictitious *Der Blaue Engel* took the place of the *Folies Bergères* in Paris. Another thriving cultural focus was Munich's Schwabing district, with artists Paul Klee and Vassily Kandinsky, and where in 1929 the violin prodigy Yehudi Menuhin made an early public appearance, aged 13. German literature, art, cinema, science, architecture, and design became trendy, rivalled only by new imports from the United States: jazz and ragtime music, and Hollywood cinema. In Berlin, Albert Einstein had become 'the pope of science'. Einstein's baffling new theories were in tune with a progressive new culture in which change for its own sake had become the theme. But not all welcomed such avant-garde developments. Modish fashion was lumped together by the reactionary Nazis and condemned as 'cultural Bolshevism'.

The Nazis gain power

There were periods of calm and stability, but with such a volatile brew, the unexpected produced the unimaginable. In 1929, the Wall Street Crash sparked what became a worldwide economic depression. In Germany, its effects were amplified when the United States called in loans which had underwritten Germany's reparations. Unable to continue these payments and already having been a victim of hyperinflation, Germany was in crisis. In the 1930 elections, the Nazi party, which only a few years before had been unknown, amassed over a hundred government representatives. It was a protest vote by a confused and dejected population. Jewish contributions to German culture continued to be acknowledged at an official level: artist Max Liebermann, a pioneer of German impressionism, became President of the Prussian Academy of Arts. But in the streets, shadowy figures appeared, smashing the windows of Jewish shops, and attacking worshippers as they left their synagogues. Virulent Nazi propaganda pilloried banks and finance, while its adherents attacked bankers and financiers.

Volatility was not confined to the stock market. Political swings are usually measured in terms of a few percentage points, but in 1932 the Nazi vote doubled. Some saw this as a warning: others dismissed it simply as yet another transient upsurge which would soon run its course and disappear. They were encouraged in November 1932 when the Nazi vote fell substantially.[14] At the same time, a new agreement drawn up at a meeting in Lausanne alleviated German's onerous reparations payments. The ageing President Hindenburg, the former Army Chief of Staff and the personification of what Germany had once been, was reluctant to let Hitler become Chancellor. But the ambitious Nazi leader, seeing what Mussolini had achieved in Italy, insisted on the job. In January 1933, Hindenburg, reassured by ministers that Hitler could be controlled, demurred and the newcomer moved onto Germany's political throne.

The spark which detonated the next political explosion came in February 1933, just before yet another round of parliamentary elections, when the *Reichstag*, the Parliament building, went up in flames. Who was responsible, nobody knew, but it gave the Nazis the chance to blame it on others. With the Reichstag building no longer available, Hitler pushed through new legislation to enable political decisions to be taken without parliamentary approval. The Reichstag Fire Decree gave the

Nazis emergency powers: 'Articles ... of the Constitution of the German Reich are suspended until further notice. It is therefore permissible to restrict the rights of personal freedom, freedom of opinion, including the freedom of the press, the freedom to organize and assemble, and the privacy of postal, telegraphic and telephonic communications. Warrants for house searches, orders for confiscations as well as restrictions on property, are also permissible beyond the legal limits otherwise prescribed.' These 'temporary' emergency powers endured. Mobs began to rampage through Berlin and other cities, beating up Jews, or people they thought were Jews, and dragging customers out of Jewish shops. Many other Germans disapproved of such barbarity, but it is difficult for individuals to confront organized mobs. Another new highly symbolic 1933 enactment was *Gleichschaltung* (synchronization), under which regional affairs became increasingly overseen by the national government.

The new parliament, with Hitler now as Chancellor, but still presided over by the by ailing Hindenburg, was formally opened on 21 March 1933. It was the beginning of a new *Deutsches Reich* (Empire), known in English as the Third Reich, the successor to the Empires of Charlemagne and Bismarck. The Nazis knew that pomp and pageantry impressed the people. Their troopers marched by drumbeat in torchlit procession through Berlin's Brandenburger Gate. But parades are not government. When a new political party comes to power, its first enactments are the most symbolic. In April, what had been prejudice now became official policy under an avalanche of new emergency rulings. Before 1933, there had been many instances of overt anti-Semitism: shop windows had been broken, posters slapped on walls, people beaten up. After 1933, Jewish shops were smashed and homes invaded by uniformed guards. Austrian impresario Max Reinhardt (born Max Goldmann, 1873–1943) was sacked as Director of the *Deutsche Theater*, which was nationalized.[15] Max Liebermann, at 86 the doyen of German art, said, 'I want to throw up more than I can eat'.[16]

But there had been nothing in writing. One of the very first enactments of the new regime was the *Gesetz zur Wiederherstellung des Berufsbeamtentums* (Law for the Restoration of the Civil Service) first published in the official *Reichsgesetzblatt* on 7 April. Even its title was cynical: what was being restored? It was one of the first of many such euphemisms: the ultimate being the 'final solution' to the 'Jewish problem' adopted in 1942. While to many it came as a bolt from the blue,[17] the new Civil Service Law had a pedigree. In his book *Judenpolitik im Dritten Reich*, Uwe Adam points to the realization that political harmony would improve if government officials could be made to sing in unison. The Civil Service had to be made politically correct, but removing unwanted elements was difficult when under law they held tenured positions. The only way to overcome such an obstacle was to change the law. The Nazis engineered machinery to override traditional administrative inertia and provide a blueprint for the 1933 Civil Service Law and its later amendments.[18] Such ideas had been mooted in the *Reichstag* as early as 1925 by Wilhelm Frick, a lawyer by training, and former head of the Munich criminal police.[19] Briefly imprisoned after the 1923 putsch, he was subsequently given a suspended sentence. He entered the *Reichstag* soon after, and became the first Nazi to hold a senior government position. In March 1933, a series of drafts for the new Civil Service law made no mention yet of the fate of Jewish officials. The main concern seemed to be a sudden bottleneck of pension payments.

Even in those first days of power, Hitler preferred his mountain lair in Berchtesgaden to Berlin. From there, late in March he demanded his ministers take direct action against Jews. The result was the *Judenboykott* on 1 April. Uniformed troopers stood outside Jewish businesses, including those of doctors, lawyers, and notaries, and urged, or sometimes forced, customers and clients to go away. Placards underlined the message. However, many ignored such calls and the boycott did not achieve its aims, in particular because it had overlooked that 1 April was a Saturday, the Jewish Sabbath, and many Jewish shops and offices were closed. Frau Margarethe Beck was at home that day and saw from her window what was going on. So did her husband. 'If you've read *Mein Kampf*, you can have no illusions,' he told her. They left Germany immediately, leaving their flat and all their belongings in the 'care' of friends that they innocently trusted. They never received a pfennig.[20]

The following day, the Nazi *Völkischer Beobachter*'s banner headlines, bold and underlined, looked like an announcement of the outbreak of war. 'The enemy of the nation warned'; 'Berlin rises up against the Jews'; 'The discipline of the aroused German People punishes abominably hateful lies'.[21] Even the newspaper's business section led with 'Jew-free stock market surge'. The *Gemeindeblatt*, the modest monthly bulletin of the German Jewish community, in April 1933 threw back the same words—'Berlin's Jewish community battles against abominable hate (*Greuelhetze*) and boycott propaganda'.

Perhaps piqued by the lack of success of the *Judenboykott*, the Nazis accelerated plans for their next moves. In Berlin, the authorities were urged to suspend Jewish schoolteachers to avoid 'unrest' in schools. On 6 April, the *Völkischer Beobachter* warned that something new was in the offing. The Civil Service was to be put on a new footing by a 'Law for the purification and preservation of the civil service'. However details were not yet fixed. Nevertheless, the next day came the first announcement of the law, signed by Frick as Minister of the Interior, by Finance Minister Schwerin von Krosigk, and by Hitler. Its Nazi-speak said that public officials recruited after November 1918 who had not undergone 'prescribed training', would lose their jobs. By 'prescribed training' and other euphemisms, the law implied that all public positions were henceforth under scrutiny. One paragraph explicitly targeted 'officials of non-Aryan descent'. This 'non-Aryan paragraph' was a last-minute amendment to the earlier drafts, but was the harbinger of what was to come later. What had been marked in the Nazi manifesto since 1920—that all Jews should be excluded from public office—was entering the statute book.

The text began by pointing out that to 'simplify the administration' certain officials would lose their jobs, and their salary would stop after three months. The first to be singled out were those who had not undergone 'the usual training' or were not 'suitable'. It was an indiscriminate statement, a catch-all excuse. However the 'non-Aryan' paragraph preferred the more euphemistic 'retirement' to outright sacking. Aryan, derived from the ancient Sanskrit word for 'noble', originally indicated peoples of proto-Indo-European origin. The word had originally been applied to languages, but in the nineteenth century, focus had shifted towards anthropology as well as linguistics, at a time when distinct racial characteristics were considered to be important. In the Nazi era, Aryan developed the special connotation of a superior Nordic/Germanic race, seen as a genetic template for the Third Reich. The hurried drafting of the initial Civil Service Law was again apparent a few days later, when on 11 April the first order for the implementation of the new law clarified as 'non-Aryan' anyone who had non-Aryan, '*particularly Jewish*', parents or grandparents. Having just one such parent or grandparent was sufficient, '*in particular*' if affiliated to the Jewish religion. Not many great-grandparents of German Jews had baptized their children. It had been too early. For a Jew to have been baptized was deemed irrelevant: parents and all grandparents would also have had to have done so. Whatever the original genetic implications, the wording of the law signalled that Jews were being singled out for special treatment. Even being a quarter Jewish was enough. The door of religious emancipation that had been opened by Bismarck in 1869 had slammed shut.

But before the law was published, a still defiant Hindenburg, reacting to pressure, had managed to persuade Hitler that Jews with war service, and those whose fathers or sons had died in the war, should be exempt from the ruling and should retain their jobs. There was also a longevity exemption for those in office since 1914. Many of those 'non-Aryans' who otherwise would have been sacked escaped through these loopholes: when the Nazis came to power, Germany had some 5,000 Jewish lawyers, making up about 20 per cent of the members of the profession. 336 out of 717 judges were temporarily exempted, as were 3,176 out of 4,585 others.[22] But many sensed that such escape routes too would be closed, and made preparations. They were right:[23] the enfeebled Hindenburg let go of his grip on life in 1934, leaving Hitler to become formal head of state, as well as political master.

Highly visible to the public, Jewish lawyers were a convenient soft target. On March 10, 1933, the Munich lawyer Michael Siegel tried to file a complaint about a fellow Jew who had been taken to the nearby Dachau concentration camp. Instead of complying, the police beat up Siegel, cut his trousers off at the knees and paraded him through the streets carrying a placard: 'I will never again complain to the police'. Such humiliation was a warning to others. Siegel later emigrated to Peru.[24]

Even before the Civil Service Law appeared in print, many guessed that something was afoot. A meeting of senior university administrators in Wiesbaden awaited the arrival of a delegate from Berlin, who had just been summoned to a ministerial meeting there, where he had been given an outline of the forthcoming edict. As a lawyer, he had been shocked. His best people were Jews. However what to do about it was a different matter, but most of those present acknowledged that if they wanted to keep their own jobs, they would have to lie down and let the Nazis have their own way.[25] On 24 March 1933, the Berlin correspondent of the London *Daily Mail* reported an exclusive interview in which a Nazi official had declared 'The Jews have abused their power in Germany. ... They are being discharged from office because they are not nationals and have failed to protect the people from Marxist infection and from atheism'.[26]

What the Nazis had in store for the Jews was not yet clear, or even decided, but the Civil Service Law and its derivatives were also the first wave of a remarkable series of ideological rulings which reflected how much importance the Nazis placed on the perceived purity of their culture. Soon after the Reichstag elections, the members of the Prussian Academy of Arts had been asked whether they would sign a declaration of loyalty to the new regime. Only nine out of 27 said they would not.[27] By this time, Liebermann had stepped down as President of the Academy, and soon resigned. Elsewhere, there came support for the new moves in a declaration by professors in universities and colleges.[28] In September 1933, another new law excluded Jews from any form of agriculture. A distinctive profile began to emerge: culture and the land itself were seen as sacred, not to be defiled.

This initial version of the Civil Service Law underwent about 15 extensions and modifications in the course of 1933 alone. Their wording was of byzantine complexity: 'The Second Ordinance on the implementation of the Law of the Restoration of the Professional Civil Service of 4 May 1933 as amended by the Regulations, amending and supplementing the Second Ordinance on the implementation of the Law of the Restoration of the Professional Civil Service of 7 July 1933 and 28 September 1933'. Obscurity was a constant feature: the initial ruling had indicated that the full picture would become clear by 30 September. In practice, a lot happened afterwards, as procedure tried to keep pace with stated objectives.

Machinery

As the legislative momentum increased, those concerned had to fill in a four-page questionnaire (*Fragebogen*) with details of name, job, war experience, and then, for each parent and grandparent, their religion (or former religion), and political affiliations. But the 'particularly Jewish' caveat in the wording of the law hinted that officials were on the lookout for Jewish ancestry. Bureaucracy had become a weapon. A single Jewish grandparent was all that was required initially to be considered Jewish. Walther Nernst, now nearly 70, haughtily chose not to reply. With two of his daughters married to Jews, some suspected that he had something more to hide, and avidly examined church registers for signs of any Nernst Jewish ancestry. There were none.[29]

A baptismal certificate, previously considered a social insurance policy, did not cancel out heredity. Jews were a race rather than a religion, although the exact implications would only be made clearer later. As the collected data came in and was analysed, the situation slowly became clearer, as subsequent actions added to the plight of those initially being placed on leave.[30] The difficulty in handling the sudden avalanche of family information, and the subsequent widening of the net from the visible Jewish community to include those with even traces of Jewish blood, and the extension of the law from academics to professionals of all kinds, uncovered many more 'Jews' than had initially been envisaged.

The genealogical input data was supplemented by information from a census of the population in 1933. People were invited to supply supplementary information in a sealed envelope, which would not remain sealed for long. Police chief Reinhard Heydrich ordered all regional police stations to register all Jews in their areas.[31] The result of all this effort was the creation of a nationwide card index (*Judenkartei*) whose objective was to list every Jew in the country,[32] a mammoth effort which was even extended to cover Jews in other countries as well.[33] More input came from denunciations by people wanting to ingratiate themselves with the authorities, and/or bearing a grudge.[34] In Berlin, Frau Hilde Walter, an acquaintance of Carl von Ossietzky, saw the implications as soon as she was instructed to fill out a *Fragebogen*, but she was soon arrested anyway after a lift-boy overheard her making a tactless remark and denounced her.[35] With eager ears everywhere, denunciations were not confined to Jews. Martin Bergsträsser of Dresden's Technische Hochschule was also overheard making political remarks. After ten months' imprisonment, and no longer suitable for an academic post, he moved to an industrial position.[36] In the face of such enthusiasm, officialdom was sometimes overwhelmed, for example with lists of those who had purchased items from Jewish shops. Not every such accusation could be taken at face value.

In total, some 1,600 scholars, teachers, assistants, librarians, etc. were sidelined in the first wave of dismissals.[37] At the beginning of 1934, United Aid for German Jewry (*Hilfsverein der deutschen Juden*) reported that of the university teachers affected by the new laws, 350 had not received any salary since October. A collection was organized. Meanwhile 200 academics had found places abroad, and 53 had managed somehow to respond to advice to 'change their occupation'. Also on the breadline were 1200 Jewish writers and journalists, 450 singers, and 150 actors.[38] In parallel with this, new legislation stigmatized the *Ostjuden*, the recent immigrants from Poland and Russia. In 1933, Interior Minister Frick instructed regional governments to stop new immigration, expel those with no residence permit, and stop their naturalization.[39] The restrictions reached some odd corners: in April 1933, Jewish boxers were excluded from the German Gymnastics League.[40]

Some several millions of public employees, together with other target groups—doctors, lawyers, students—had to find proof of being Aryan, and 'particularly' of not having Jewish blood. All over the country, thousands of priests, local government officials, and archivists were conscripted into a vast new bureaucratic machine to furnish citizens with an *Abstammnachweis* or *Ariernachweis*—a proof of being Aryan—a genetic/cultural passport. Everyone subject to Aryan rulings was expected to have a *Nachweis*. If Aryanship were doubtful, documentation had to be sought from the Ministry of the Interior's department for *Sippenforschung* (race investigation).

Another law told Jewish lawyers that their licence to practise was being revoked, but initially included exemptions for war service, so that many of them were able to continue working. Among highly qualified professionals, doctors are among the most visible to the general public. As such, they were convenient targets for official public persecution. In 1933, some 13 per cent of German doctors were Jewish,[41] a number out of all proportion to the percentage of Jews in the population as a whole (less than one per cent on average, but concentrated in major cities, with 1.75 per cent in Hamburg).[42] In parallel with the legislation for the Civil Service Law, German doctors received a short 'Dear Colleague' letter from the German Doctors' Association (*Hartmannsbund*). Enclosed was a questionnaire on which each recipient had to fill in personal details, and those of parents and grandparents, together with those of their spouse. They were warned that supplying 'negligent' information would be severely punished. It looked to have been hastily organized: as with the Civil Service Law, additional conditions appeared on a week-by-week basis. Even while the doctors were filling in their questionnaires, the implications were made clearer: a new law said that non-Aryan doctors would lose the right to work for social security organizations. However, the same initial exemptions were granted: those who had been soldiers, or had served in field hospitals, or whose fathers or sons had died in the war.[43] Harassed, about one third of the Jewish doctors left Germany in 1933 alone;[44] 32 went to Shanghai. To tighten the noose, in January 1935, non-Aryans were banned from becoming doctors, dentists, or pharmacists, and could not sit the relevant entrance examinations.[45] By 1939, 2600 doctors would be dismissed from the profession.[46] Aryans would ensure the health and well-being of German nationals.

In an eerie echo of the Russian May Laws of 1882, Jews were excluded from occupations where they were allegedly 'overrepresented', in particular lawyers and judges. Another edict limited Jewish access to schools and universities. On 25 April 1933, the Law against the Overcrowding (*Überfüllung*) of German Schools and Institutes of Higher Learning stipulated that the number of non-Aryan Germans who may be admitted to schools, colleges, and universities, could not exceed a level comparable to the percentage of non-Aryans in the German population as a whole. According to data compiled by Jewish self-help organizations, the Jewish student population in Germany as a whole was 2.7 per cent for men, (3.3 per cent in Prussia), even higher for women. The same data showed that the redefined Jewish population of Germany as a whole was 0.9 per cent.[47]

Emigrants and victims

Doctors were more discernible to the population as a whole than university professors, and some attacks on them were particularly vengeful. Their contributions to the advance of medicine were imperiously attacked, their names expunged from medical records, and credit for their achievements erased. Josef Löbel was targeted from two directions. Born in Franzensbad (Bohemia) in 1882, he had a successful career both as a hydrotherapist and as a writer of popular science. His literary work included novels and a standard biography of Robert Koch, one of the pioneers of modern bacteriology, translated into more than a dozen languages. Before the Nazis came to take him away in 1938, he committed suicide in his bath after taking drugs. His wife died in the Theresienstadt camp.

Ismar Boas, born in 1858 and one of the pioneers of modern gastroenterology, had left his professorial post in Berlin and was an old man when the Nazis came to power. Forced to sign away credit for his published work and with his private practice threatened, in 1936 he moved to Vienna. When the Nazis appeared there in 1938, he committed suicide by taking barbiturates. Another Vienna doctor who took his life was Walther Hausman, born in 1877 who extolled the curative properties of light. Dismissed from his various posts as a Jew, he hanged himself. Josef Jadassohn, born in Liegnitz in 1863, was also an old man in 1933, but had had a distinguished career as a dermatologist with a list of skin conditions named after him. His emeritus position in Breslau was annulled, and his name erased from title pages of specialist papers. Crushed, he emigrated to Switzerland, where he died in 1936.[48] Such dismissals soon led to the appearance of 'cultural ghettos': German Jewish schools, German Jewish theatre, cinema, opera, art . . . ,[49] with its own *Kulturbund*.

The definitions of 'non-Aryan' in the 1933 Civil Service Law deemed that a person with one Jewish grandparent was culturally undesirable. Other than the strict selection procedures used for Hitler's SS protection guard, and before the promulgation of the Nuremberg Laws, this was the widest definition used, and displayed the importance the Nazis placed on arts, science, and the humanities. In filling in the necessary forms and obtaining the necessary papers, some were surprised to discover that they had a Jewish grandparent and were thus *persona non grata*. If they had not known, then most of the time neither did anyone else. However some observant colleagues guessed. Wolfgang Pauli, one of the pioneers of modern quantum theory, was brought up in an emancipated intellectual family in Vienna and was totally ignorant of his Jewish blood until as a student at Munich he met X-ray crystallography pioneer Paul Ewald, then a lecturer. Ewald relates[50] that Pauli, then 19, made some remarks 'which showed he had not the slightest idea he was Jewish'.

EWALD: 'Surely you are Jewish?'
PAULI: 'Nobody has ever told me that and I don't believe that I am'.

When Pauli returned home for a vacation, he asked questions and found out. For his part, Ewald had a distinguished career. Born in Berlin in 1888, in 1921 he had inherited Erwin Schrödinger's chair at Stuttgart Technical University (before Schrödinger became famous), and eventually rose to be rector of the establishment. However he did not see eye to eye with Stuttgart's Nazi establishment. In addition he had Jewish ancestry himself and had a Jewish wife. Without waiting for any decision concerning his fate, he resigned from his post as Rector at Stuttgart TU, and left for the UK in 1937, after 'voluntary retirement', moving first to Cambridge, then settling in Belfast.

The situation of Jews in the 1930s is reflected by the experience of the nuclear physicist Victor Weisskopf, born into a liberal Viennese Jewish family in 1908. He became a victim of Nazi race laws and went on to play a major role in the development of the Atomic Bomb. Weisskopf was aware of his Jewish background: he knew that his father, Emil, had as a child helped his own father, the village slaughterer, in his work in a ghetto village. Emil Weisskopf's academic potential had been recognized by an enlightened local rabbi, who made sure the boy was schooled beyond the confines of the ghetto, an important first step towards a successful career as a corporate lawyer. One generation later, the even more assimilated Victor Weisskopf related '[being Jewish] did not mean very much to me. At most, I felt some pride about belonging to a group that emphasized intellectual values'.[51] However, the first time Weisskopf visited a synagogue was when he visited Israel in 1956 (at the age of 48). He never had the customary *barmitzvah* as a boy, and was not circumcised. In Berlin in 1931 he encountered Nazis and anti-Semitism, but relates that this had not made him feel any more Jewish. Instead, he felt offended more as a fervent internationalist who 'abhorred nationalist tendencies', attributing this feeling to the continual endeavour of the dynamic Jewish community in Vienna to assimilate.

Implementing Nazi racial research policy was Achim Gercke, who as a 22-year-old science student had dutifully begun in 1925 to compile a secret register of German university professors with Jewish blood. It would be a 'weapon' to cleanse the German race of the last 'recognizable Hebrew' and expel them from the country.[52] His lists focused particularly on Göttingen. As a reward for such dedicated work, Gercke became a Nazi official, and later a consultant expert on racial matters for the Ministry of the Interior. However, in January 1935 he was arrested as a homosexual (then considered a criminal offence), excluded from the party, and placed in a 'probation battalion' of soldiers who had to 'restore their military honour'. He was fortunate, as voices in the SS called for the death penalty for homosexuals: some were sent to concentration camps, where they were identified by a distinctive pink triangle badge.

Gercke's rants provide a useful mirror of contemporary feeling. Writing on 'The Solution of the Jewish Question' in the *Nazionalsozialistische Monatshefte* in 1933, he underlined that the Jews as a 'community of blood' require some special racial treatment. '*Everyone* agrees that the current situation is untenable: uncontrolled development and the equal treatment of Jews have led to a situation where the Jews are able to exploit the competition in an unfair way, resulting in the loss of positions important to the German people. Plans and programmes must be far-sighted and not just attempts to solve a momentary inconvenience. ... Only then can *Ahasver* (the wandering Jew) be forced to take up his staff for the last time'.[53] In 1933, the idea of a 'Final Solution' was beyond the horizon.

The sheer speed of the new laws caught many unprepared. Hans Krebs (1900–81), born in Lower Saxony and with a Jewish mother, had worked as a doctor in Hamburg and a researcher in Freiburg, where he had made important discoveries on metabolic mechanisms, earning him official recognition for 'outstanding scientific ability'.[54] Just a few months later, a letter from the same authority told him bluntly that his services were no longer required. Krebs quickly transferred to Cambridge. In 1953 he was awarded the Nobel Prize for Medicine for his work on metabolic chemistry (with Fritz Lipmann). In all, some 30 Nobel prizewinners left Germany before 1939. Some of them had the award even before they left; some were in the middle of their careers; some were too young even to know in which direction their future careers lay; some of their children would also become Nobel laureates.

Many of the victims of the new laws had colleagues and admirers overseas who were also shocked by the suddenness and the implications of the new moves. The reputation of research scientists like Krebs was not restricted to the institute, or even country, in which they worked. In parallel, other anti-Jewish moves, such as the official *Judenboykott* on 1 April, had been widely criticized overseas, particularly in the USA, and threatened a commercial backlash that could endanger Germany's still fragile economy. The Nazis noted that they could not always be seen to do as they liked.

The removal of Jewish university teachers helped ensure that future generations of students would receive 'suitable' instruction, but these students still needed to consult and read 'suitable' material. Two months after the fire that destroyed the Reichstag, more fires were lit.

This time it was books, huge piles of them, by Bertold Brecht, Albert Einstein, Sigmund Freud, Heinrich Heine, Alfred Kerr, Thomas Mann, Erich Maria Remarque ... even Erich Kästner, author of the bland *Emil and the Detectives*, was deemed to be 'against the German spirit'. Anything deemed to have Jewish, communist, or pacifist leanings was taken off the shelves and trashed. In the library catalogues, the entries were carefully obliterated, and the books fuelled ceremonial bonfires in several major cities. The event on the Göttingen campus was particularly well orchestrated, with the Nazi student *Korps* playing a prominent role, but the biggest bonfire was in Berlin, where an initial procession filed through the Brandenburger Gate and along *Unter den Linden* before assembling. The fire brigade was on hand, but this time in a different role—with cans of petrol in case the books didn't burn fast enough. Crowds of onlookers cheered and munched sausages. Some scrambled up lampposts or perched on roofs to get a better view.

Contemporary German culture was expurgated. Thomas Mann, born in Lübeck in 1875, achieved fame with his epic *Buddenbrooks*, published in 1905, and went on to produce a series of literary masterpieces, including *Death in Venice* and *The Magic Mountain*, and in 1929 was awarded the Nobel Prize for Literature. A fervent republican, in the 1920s he continually spoke out against Nazism. He was on vacation in Switzerland when they came to power in 1933, and chose not to return to Germany. Although his books were initially not among those consumed by the 1933 bonfires, his citizenship was revoked in 1936. His elder brother Heinrich, author of the book which became the successful film *Der Blaue Engel* starring Marlene Dietrich, left the country before the Nazis came to power. Another refugee writer was Bertold Brecht, born in Augsburg in 1898. He became enamoured with Marxism in the 1920s, at the same time as his *Threepenny Opera* became a major hit in Berlin. In 1933 he fled Germany, eventually settling in the USA in 1941; but even there his openly pro-communist stance forced him to leave, in 1947.

Alfred Döblin, born in Stettin in 1878, was many things, as well as a Jew. After studying medicine, he became a practising psychiatrist in Berlin. In addition, he was an influential journalist and author. His books, including the futurist *Berlin-Alexanderplatz*, became landmarks of contemporary literature. However for the Nazis, such progressive writing was synonymous with contamination. Döblin was a marked man. Despite his war service record, in 1933 he pre-empted Nazi moves and left Germany via Switzerland, eventually moving to Paris, where he became a French citizen. During the war he emigrated to the USA and began another career as a Hollywood scriptwriter. His son Wolfgang, born in Berlin in 1915, was a promising mathematician who nevertheless had to join the French Army in 1938. In 1940, following the German invasion, he was awarded the *croix de guerre* for bravery, but committed suicide after his regiment surrendered. While a soldier, Wolfgang Döblin had continued with his mathematics and sent a letter to the French Academy of Science, which remained unopened until 2000. His contribution to the theory of probability threw new light on the celebrated Kolmogorov Equation.

Unlike literature or science, the language of music needs no translation, and musicians have fewer obstacles to emigration. Here, the figurehead was Arnold Schönberg, born in a traditionally Jewish neighbourhood in Vienna in 1874, whose modernist atonal compositions were to twentieth-century concert music what relativity and quantum theory were to its physics. The emergence of this music in the concert hall also paralleled that of jazz on the popular scene. As the music primarily of African Americans, jazz was evidently non-Aryan. But the Nazis lumped atonal music with jazz as degenerate. Although Schönberg had converted to Protestant Christianity, and had been appointed Director of the Prussian Academy of Arts, both his music and his background classified him as undesirable. After leaving Germany in 1933, he reaffirmed his Jewishness and initially headed for Paris before finally settling in the USA. Another emigrant was Billy Wilder, born Samuel Wilder in Sucha Beskidza, Poland, in 1906, who moved with his family to Vienna, where he became a journalist. Transferring to Berlin, he shifted the focus of his writing from newspapers to the cinema screen. In 1933, he moved to Paris and made one film before moving on to Hollywood, where he went on to earn Academy Awards for producing, writing, and directing. He died in Beverley Hills in 2002. Art too was targeted, and the prejudice culminated in a specially organized exhibition *Entartete Kunst* ('Degenerate Art') held in Munich in 1937. The publicity billed it as the work of 'lasciviously motivated no-hopers' who had been 'crowned' by 'Jewish cliques', and invited the public to judge for themselves. Ironically, it attracted far more visitors than a parallel 'Great German Art Exhibition'. After the annexation of Austria in 1938, the exhibition was moved to Vienna.

After the various 1933 directives and some carefully arranged Jew-baiting had done their work, the Nazi press crowed 'They have gone!'.[55]

At that time, 'they' did not mean Jews as a whole, but merely those in important professional positions. Hermann Göring, widely seen as Hitler's deputy, stressed that an initial objective had been attained, claiming 'the decent Jewish merchant who is willing to stay in Germany as an alien, protected under the law, will be allowed to pursue his business undisturbed'.[56] Göring understood; his family had some big Jewish skeletons in its cupboard.[57]

Culture had been purged, but, despite the boycott of Jewish business, by 1935 Germany had not yet achieved the prosperity that the Nazi manifesto had promised. There could be only one reason—the continuing presence of undesirables, 'particularly' Jews. It was time to increase the stakes. The Nazis held vast annual rallies in the Bavarian town of Nuremberg, and the 1935 event provided a platform for intensified Jewish oppression. On 15 September, after some last minute pressure by Hitler, redolent of the lead-up to the 1933 Civil Service Law, came a new set of directives. With these, Nazi oppression of the Jews was no longer confined to culture and the civil service: the Jews were relegated to second-class citizens, social outcasts. The Nuremberg decrees—the obliteration of almost a century of emancipation—make an even more stark impression in today's climate of 'political correctness'. One of the new laws, 'to protect German blood and honour', prohibited marriage (even if it had been formalized outside Germany), and even sexual relations, between Germans and Jews. German women under 45 were no longer allowed to work in Jewish households. Jews were forbidden to show the German flag. The penalty was imprisonment or hard labour. The evolved definition of a Jew had by now become remarkably convoluted: for example anyone born after 31 July 1936 as a result of extra-marital relations with a categorized Jew, and whose parents were still unmarried, technically became a Jew.[58] A supplementary citizenship law stripped Jews of their nationality, and their papers were stamped with a large 'J'. Amid a scramble for Aryan certificates, the emigration exodus accelerated.

The Nuremberg Laws introduced a new category, that of a '*Mischling*', or crossbreed. The draft classification scheme for the Nuremberg Laws had been very flexible, leaving the final decision to Hitler himself.[59] The final selection criteria under which *Mischlinge* with one or two Jewish grandparents were deemed to be Jews were based on choices which the subjects had made personally—that they were a member of a Jewish congregation, or had married a Jew. Only in marginal circumstances were baptismal certificates of any use. While meant to provide clear definitions, these rulings instead opened up the grey area of marriages between *Mischlinge*. Another category was *Mischehe*, mixed marriages. Those married to non-Aryans were also excluded from civil service professions. Registry officials were discreetly told not to carry out such marriages.[60] In 1938, a law for the 'unification' of marriage and divorce facilitated the annulment of mixed marriages. In such a climate, those who had married Jews felt particularly vulnerable to the whims of Nazi jurisdiction. The Nuremberg Laws also affected those who had Jewish spouses. Karl Meissner had a distinguished career in the study of atomic spectra, becoming director of Frankfurt's Physics Institute. In 1937 he lost his post because of his Jewish wife, and soon moved to the USA, but his wife died in 1939.

Motivated particularly by Einstein's work and achievements, and by the attraction of the recondite quantum world, Jewish talent had been attracted wholesale to mathematics and exact science. These areas were especially hard hit by the new laws, and the departure of a whole community changed the profile of German science. When warned by IG Farben founder and 1931 Nobel Chemistry laureate, Carl Bosch that sacking so much talent was ruining German science, Hitler retorted 'If the dismissal of Jewish scientists means the annihilation of German science, then we shall do without science for a few years'.[61]

In *Mein Kampf*, Hitler alleged that scientific and technical education was at the root of 'cowardly lack of will'.[62] Hitler's disdain for science was underlined by the appointment of a former schoolteacher, Bernhard Rust, as Minister for Science and Education, further antagonizing established scientists like Planck. But others greeted the various 1933 rulings as a *nationaler Aufbruch* ('national awakening'), especially when new money started to pour into the coffers of the Kaiser Wilhelm Society. However the Kaiser Wilhelm Institute for Physical Chemistry in Berlin, led by Fritz Haber, was hamstrung. Carl Neuberg, Director of the Biochemistry Institute, was forced to leave in 1934 and was replaced by Adolf Butenandt. Another Jewish Director to lose his position was Max Bergmann, who in 1921 had been appointed to lead the new Leather Research Institute. In 1933 he left for the Rockefeller Institute for Medical Research, New York.

Other Kaiser Wilhelm institutes carried on their business relatively unperturbed. A new Institute for (Plant) Breeding Research had no Jews at all on its payroll. Race and heredity studies, together with rearmament, appeared as new scientific objectives. Another goal was self-sufficiency in raw materials, but this was hardly new: Haber's invention of ammonia synthesis during the First World War had been a major step in that direction. In the aviation sector, the Jewish pioneer aeronautical engineer Theodor von Karman left, but a young enthusiast called Wernher von Braun was studying at Berlin's Technical University. As well as compensating for the loss of von Karman to Germany, he would later go on to lead the US space effort.

The banished intellectuals were only the tip of the Jewish iceberg. With their history punctuated by expulsions and emigrations, those in Germany could now see what was coming. Some 35,000 fled the country in 1933, mostly going elsewhere in Western Europe, but some going to the Americas. Among the onlookers at the Berlin book-burnings in 1933 had been some of the authors themselves. They, and others, had wondered what would happen next. The caprice of a minority was becoming the mood of the majority. In 1933 it was books that went up in smoke, A few years later, it would be shops and synagogues, starkly symbolized by the *Kristallnacht*, the night of broken glass, in November 1938. Where thousands of books had been burned on Berlin's Opernplatz (now Bebelplatz) in 1933, lines from Heinrich Heine's 1821 play *Almansor* are today engraved as a grim reminder: '*Dort, wo man Bücher verbrennt, verbrennt man am Ende auch Menschen*'. ('Where they burn books, they ultimately also burn people').

NOTES

1. Brustein, p.207.
2. Gilbert, p.541.
3. The size of the German Army would soon be limited by the terms of the armistice.
4. Gilbert, p.505.
5. Brustein, p.107.
6. Brustein, p.289.
7. Elon, p.349.
8. Gilbert, p.523.
9. Brustein, p.220.
10. Elon, p.229.
11. Gilbert, p.500.
12. Peierls, p.12.

13. Elon, p.232.
14. McDonough, p.19.
15. Friedländer, p.12.
16. Elon, p.401.
17. except Einstein and Leo Szilard.
18. Strauss, Immigrants, p.72–3.
19. Friedländer, p.28.
20. Wiener Library, Eyewitness Accounts, pIIf, No.1084.
21. *'Die Disziplin des erwachten deutschen Volkes straft die Greuelhetze Lügen'.*
22. Evans, p.437.
23. Friedländer, p.145.
24. http://www.haaretz.com/weekend/week-s-end/coming-full-circle-1. 347088.
25. AIP oral history P. Ewald 4596.
26. Quoted in Simpson, J. *Unreliable Sources: How the 20th Century was Reported*, London, Macmillan, 2010, p.217.
27. Friedländer, p.11.
28. Noakes and Pridham, p.250.
29. Mendelssohn, p.105.
30. Letter from Selig Hecht to a US colleague, 20 June 1933, Wiener Archive, London, file 1527.
31. Adam, p.110.
32. Jutta Wietog: *Volkszählungen unter dem Nationalsozialismus – eine Dokumentation zur Bevölkerungsstatistik im Dritten Reich.* Berlin 2001.
33. Friedländer, p.199.
34. Evans, p.438.
35. Wiener Library, Eyewitness Accounts, PIIa No.20, 1090.
36. SPSL Box 324.
37. Noakes and Pridham, pp.249–50.
38. Wiener Library, London, file 602/8.
39. Friedländer, p.27.
40. Evans, p.438.
41. (estimates vary between about 7000 and 9000), according to Niederland.
42. Zentral Ausschuss der Deutschen Juden für Hilfe und Aufbau, Wiener Library files 602/6/1–7.
43. *Der Spiegel*, Vol. 48, 1979, p.249, citing Leibfried, L. and Tennstedt, F., *Berufsverbote and Sozialpolitik 1933*, Univ. Bremen, 1980.
44. *Statistik des Deutschen Reichs*, 1936, cited in Niederland.
45. Adam, p.84.
46. Cornwell, p.155.
47. Zentral Ausschuss der Deutschen Juden für Hilfe und Aufbau, Wiener Library files 602/6/1–7.

48. These case histories and others are listed in *Autobiografie und Wissenschaftlicher Biografik*, ed. Kron, C.D., Text und Kritik, Munich, 2005, Vol. 23, p.74.
49. Schwersenz, J., *Die Versteckte Gruppe*, Wiohern, Berlin, 1988, p.24.
50. AIP oral history 4523.
51. Weisskopf, p.203.
52. Greenspan, p.137, cited in Farmelo, p.131.
53. *Nazionalsozialistische Monatshefte,* Vol. 4 (1933), p.195–7, translated in *Jewish Immigrants of the Nazi Period in the USA*, Strauss, H. (ed), Vol. 4, *Jewish Emigration from Germany*, 1933–42.
54. Krebs, H. and Martin, A., *Reminiscences and Reflections*, OUP, Oxford, 1981.
55. Elon, p.397.
56. Lindemann, p.523.
57. Deighton, L., *Fighter, the True Story of the Battle of Britain,* Jonathan Cape, London, 1977, Hermann Göring, p.12.
58. Adam, p.101.
59. Friedländer, p.148.
60. Strauss, Immigrants, pp.72–3.
61. Mendelssohn, p.146, and Beyerchen, A., 'Anti-intellectualism and the cultural decapitation of Germany.' In *The Muses Flee Hitler: Cultural Transfer and Adaptation, 1930–1945,* Jackman, J.C. and Borden, C.M. (eds) pp.79–91. Washington, DC: Smithsonian Institution Press, 1983.
62. Cornwell, p.31.

4 Emblematic emigrants

The influential British biologist J.B.S. Haldane once described Albert Einstein as the 'greatest Jew since Jesus'.[1] Einstein's work released a tide of scientific discovery which changed our view of the world, and propelled him to the front of the public stage. This visibility made him both an icon to be admired, and a ready target for attack. From his elevated perch, he was able to anticipate what caught others by surprise. Throughout his life, he was highly mobile, chasing admission to schools and universities, later seeking teaching jobs, and finally progressing through a series of increasingly prestigious university posts. Twice in his life he made prescient decisions to emigrate that went far beyond his immediate career advancement.

Although Einstein is probably the most well-documented scientist of the twentieth century, his role as an influence on and an inspiration for other scientists is less well known. He was born in Ulm, Württemberg, in what is now Germany, in 1879 into a family of non-practising, assimilated Jews who had lived in that part of the world for hundreds of years. Although his parents were married in a synagogue,[2] they had discarded many of the trappings of their religion. The family had retained traditional biblical first names well into the nineteenth century, but 'Albert' was a further sign of assimilation. His father was an entrepreneur, trying his hand at whatever he thought worth exploring, wherever he could explore it, dragging his long-suffering family behind him. It was a precarious way of making a living. After several ventures, Einstein's father noticed a demand for electrical expertise, and in 1890 moved to Munich, installing equipment for power stations.[3]

At the age of six, young Einstein was introduced to the violin. Although initially he found this a chore, he persevered to become a skilful player, a useful social accomplishment. He attended local schools, where, under Bavarian law, all pupils had to have religious instruction. To counterbalance Christian conditioning, he had Jewish religious instruction at home. Although he did not undergo the ancient rite of *barmitzvah* (confirmation) at the age of thirteen, and could not read Hebrew, he became religious while a teenager. This did not last, but he later returned to his Jewish roots and became a fervent supporter of Judaism and Jewish culture.

Einstein's father, always looking for new opportunities, moved to Italy in 1894. While the family was used to having to accept such fortuity, Einstein senior did not want this to interfere with his son's schooling, so Albert did not follow the rest of the family. Stranded in Munich, he felt rootless and abandoned, and dropped out of school. But with his distant father continually pushing, he followed the family's nomadic tradition and transferred to Switzerland to complete his schooling and embark on a university education. But he had other reasons for making this move.

Einstein was born a citizen of the kingdom of Württemberg, which had recently become part of Bismarck's new German Empire. Although the state retained a certain level of autonomy, its army was under Prussian control. Conscription had been introduced in France during the Revolution and had provided the manpower for Napoleon's vast armies. Prussia followed the French example and adopted conscription to boost its level of military preparedness. Highly principled even when young, Einstein did not want to incur any kind of obligation for military service. By leaving the country, he deliberately lost his Württemberg/German nationality. Under the rules of the time, if a boy left the German Empire before the age of seventeen, he was no longer liable for military service. The young Einstein was left stateless for the next five years, but this was less of a problem then than it is now.

In 1896, he began studies at what is now the Swiss Federal Technology Institute (*Eidgenössische Technische Hochschule*) in Zurich. As well as learning about science, he also met two students who were to change his life: one became his first wife, the other was a gifted mathematician—Marcel Grossmann, born into a Swiss–Jewish emigrant family in Budapest in 1878. Equipped with a modest Zurich diploma, Einstein now had to make a living. However, one immediate requirement was to become a Swiss citizen. Einstein always paid detailed attention to such matters, and had got his application under way while still a student. In 1901, he duly became Swiss, and was to use this passport for a long time. However on doing so, he again became liable for military service, which he had gone to such lengths to avoid in Germany. This time he was deemed unsuitable due to flat feet and varicose veins.[4]

Then, just as now, an undistinguished university education is no sure stepping stone to career success. After a precarious series of temporary schoolteaching positions, on the recommendation of Marcel Grossmann's influential father, Einstein became a humble administrator in the Swiss Patent Office in Berne. There, in his spare time and alone, he thought through some of the most important scientific problems of the time, and wrote down his findings in a series of papers, which appeared in 1905. His special theory of relativity, which dictated what follows when light always travels with the same velocity, and the quantum implications of the photoelectric effect were just two of the subjects he treated.

But there was no immediate acclaim. Einstein's own obscurity and the difficulty, if not impenetrability, of the concepts he introduced meant that his ideas initially fell on unfertile ground. Today, it is inconceivable that such breakthroughs could be made by an amateur scientist. Even a hundred years ago, it was remarkable that such a modest and anonymous figure could get so much work published at all. But slowly the importance of these contributions was realized, and Einstein moved upwards through a series of university posts, first in Berne, then in Zurich, where his friend Marcel Grossmann, recently appointed professor of mathematics, had influence. Wheels started to turn and Einstein was invited to join him in a new professorial post for theoretical physics.[5]

But the newly created job did not pay well, and offers were beginning to come in from other directions. For a brief sojourn in Prague, then part of Austria-Hungary, Einstein was required to state his religion. Writing 'none' was not acceptable, so he put '*Mosaisch*', the creed of Moses, stressing his closeness to Judaism, but not necessarily his full allegiance. Einstein then returned to Zurich, where the authorities had noted that he lacked the 'unpleasant peculiarities of character' of other Jews.[6] As awareness of his work grew, in 1913 Max Planck and Walther Nernst travelled to Zurich to see if Einstein would be interested in moving to Berlin. As well as a professorial chair with no obligation to teach, they carried several other pieces of bait: membership of the prestigious Prussian Academy of Science, and directorship of a new institute, later the Kaiser Wilhelm Institute for Physics, established in 1917. Against the wishes of his wife, in 1914 Einstein moved back to the country of his birth, and as a university professor reacquired the German nationality which he had once taken pains to lose.

Einstein looked forward to a new period of scientific productivity. The ideas for what would soon become his greatest contribution to science were crystallizing in his mind. But arriving in the German capital in 1914, he was surprised by the bellicose patriotic fever that gripped the country. Having earlier severed his ties to what was after all his homeland, and having lived in neutral Switzerland for almost two decades, he had not realized that Germany and the rest of Europe were on the brink of war. Hot-headed generals and young hawks alike seemed to relish the prospect of war, but a few knew otherwise. A new representative in Britain's parliament, Winston Churchill, warned that such small professional armies were obsolete, and that when 'mighty populations are impelled on each other … the wars of peoples will be more terrible that those of kings.' The end result would be 'the ruin of the vanquished and the scarcely less fatal dislocation and exhaustion of the conquerors'.[7] Despite such warnings, Europe continued on its inexorable collision course. After the spark came on 28 June 1914, there were scenes of public jubilation that these days are seen only after major national sporting successes. But Einstein had just made his first address to the Prussian Academy of Science and had other things on his mind.

In spite of his pacifist leanings, Einstein's pride in his new German nationality was piqued by accusations of treachery and cruelty by German armies. In 1914, Einstein was one of the 93 signatories of a widely published manifesto which accused the Allies of having 'unleashed Mongols and Negroes against the white race'.[8] Later, he regretted how impulsive he had been, and was one of three signatories of an obscure counter-manifesto. A more visible platform for pacifism and pan-European collaboration was the *Bund Neues Vaterland* (New Fatherland League), of which he was a founder member. Pushing for peace while Germany was overrunning more territory, the league was out of step with public sentiment and was outlawed by the German authorities. Nevertheless, Einstein boldly continued to criticize militarism, to advocate peace, and to collaborate with international pacifists. Initially this did not appear to hinder his career, and in 1916 he succeeded Max Planck as President of the German Physical Society. His criticism of war and its supporters became laced with a new dimension: 'Ostentatious Teutonic muscle-flexing runs rather against my grain.

I prefer to string along with my compatriot Jesus Christ. … Suffering is more acceptable to me than resort to violence'.[9] (By 'compatriot' he was alluding to Christ's Jewishness, not to any Christian beliefs of his own.) However, not all pacifists were content with making speeches. Others turned to direct action. One, Friedrich Adler, shot Austrian Prime Minister Karl von Stürgkh, who earlier, as Minister for Education, had been influential in bringing Einstein to Prague.

As the war raged around the frontiers of Germany, and as Fritz Haber and his team produced new weaponry with poison gas, Einstein built his most majestic contribution to science. The ideas for his General Theory of Relativity were so radical that he found it difficult to express them in the mathematical form needed to provide a framework for calculation. He got help from his loyal friend Marcel Grossmann, whose mathematical speciality at Zurich was modern geometry; mathematics forged in Göttingen half a century before by Riemann. Several times in his life, Einstein acknowledged the debt to his friend from student days. But the effort of this epic work amid the upheaval of war seems to have taken its toll on Einstein. Germany struggled with food shortages, but Einstein, who still retained his Swiss citizenship, was helped by occasional food parcels from that neutral country. Bedridden in 1917 with what was eventually diagnosed as a stomach ulcer, Einstein moved to be near his cousin, Elsa, who lived in Berlin and who cared for him during his sickness. They eventually married in 1919, after Einstein obtained a divorce from his first wife, under conditions which included a prescient proviso under which she would eventually receive any Nobel Prize money.

The 1919 solar eclipse observation confirmed General Relativity and cemented Einstein's position as the ultimate authority of the incomprehensible. Besieged by reporters and followed by cameramen wherever he went, he became a celebrity of his era, called upon to give speeches and to comment on all matters of the moment. For some fourteen years, until he retired to the monastic quiet of Princeton in 1933, Einstein spent a major part of his life in the public arena—giving lectures on science, and acting as a spokesman on political matters well beyond his immediate control—supporting Zionism, and pleading for a conservative Germany. His visibility and achievements made him a hero for other Jews, a role-model for aspiring scientists, but also a convenient target for those with a grudge. Einstein was often portrayed as the personification of the 'mad scientist', oblivious to everything but his own work. But in fact he was highly perceptive, and could discern trends before others. His ascendance in the German public arena also coincided with the emergence of the Nazi Party as a political force: when he first became famous, the Nazis were an obscure Munich underground group; when they finally came to power, Einstein had gone. He did not wait for the 1933 Civil Service Law.

As well as his own milestone contributions, Einstein had an enormous impact on science through his presence, which seemed to inspire all who came into contact with him, despite his lack of enthusiasm for teaching. Einstein frequently said that teaching and the preparation of lectures interfered with his train of thought.[10] By the time he was wooed by Berlin, he had had enough of giving classes, which 'got on his nerves', and welcomed the prospect of a post which was both prestigious and lecture-free.[11] From 1914 to 1932, he was an inspiration to a new generation of German-speaking scientists. Eugene Wigner, later to earn the Nobel Physics Prize, in 1963, wrote of the regular Wednesday afternoon colloquia of the German Physical Society in 1920s Berlin, where the front row of seats was traditionally reserved for important names: Max Planck, Max von Laue, Walther Nernst ... 'At the head of it all was Albert Einstein. ... Most great men are respected, but Einstein also inspired real affection'.[12] 'Einstein's clarity of thought and skill in exposition were matched with a simplicity and an innate modesty. He did not want to intimidate anyone'. After being impressed and motivated, Wigner describes how coming into contact with Einstein could also help in the search for jobs. Otto Frisch also mentions how impressed he was to see and hear Einstein.[13] Never one to shrink from opportunity, a young student named Leo Szilard became Einstein's scientific stalker.

In Berlin, Einstein had been reminded of the Jewish heritage of which he had first become aware as a teenager. 'I saw worthy Jews basely caricatured. ... I saw how schools, comic papers, and innumerable other forces of the gentile majority undermined the confidence even of the best of my fellow-Jews'.[14] As Einstein's visibility increased, his numerous appearances in public had made him a target for political hotheads. In 1920, rowdy right-wing elements broke up an Einstein lecture at Berlin University. It seemed to catch him by surprise, and the Minister of Culture formally apologized. But Einstein had taken note and would not be surprised again. Soon after, a bewildered group calling itself 'German Scientists for the Preservation of Pure Science' held a major public meeting to denounce the incomprehensibility of relativity. Einstein himself attended, and subsequently wrote to newspapers to vent his anger and indignation. As he had done in 1914, he later regretted having been so impetuous, and resolved to be more equivocal. But this was not easy amid continuing public acclaim on one hand and increasing criticism on the other. In 1921, he made his first of many trips to the United States, a country to which he was increasingly drawn.

Einstein's first transatlantic visit was with Chaim Weizmann, head of the World Zionist Organization and later to become the first President of the new nation of Israel. Their primary objective was to raise money for the Zionist cause, with Einstein in particular focusing on funding for a Jewish university in Jerusalem. Einstein had been astutely targeted by Zionists and their careful efforts were soon rewarded. It was as though the incongruity of Einstein's European status needed to be compensated. Enigmatically, he said 'I am against nationalism but in favour of Zionism'.[15] In a public speech during his first visit to Palestine in 1923 he said 'This is a great age, the liberation of the Jewish soul'.[16]

A growing awareness of currents deep beneath the surface of Germany society had been sharpened for Einstein by his own experience and by the highly symbolic assassinations of activists like Rosa Luxemburg. But in 1922, the murder of Foreign Minister Walther Rathenau alarmed him. 'I had not anticipated that hate, delusion and ingratitude could go that far', he wrote in an obituary notice for his friend.[17] Later that year, Einstein was invited to speak at the centennial meeting of the German Association of Scientists and Physicians. With rumours that he was on the Rathenau murderers' hit list, he preferred instead to go on a world tour. En route, he learned that he had been awarded the Nobel Prize. Einstein expertly rode a fresh wave of popularity. In his continuing travels he was introduced in India to the poet Tagore and to Mahatma Gandhi; in the USA, he met Charlie Chaplin (who invited him to the world première of his new film, *City Lights*)[18] and John D. Rockefeller. In Britain he met H.G.Wells and George Bernard Shaw.

Away from this razzmatazz, he kept a firm eye on new opportunities in the United States, which was eager to promote an awareness of culture to underline its brash new prosperity. One offer was from the newly established California Institute of Technology (Caltech) in Pasadena. With its attractive Mediterranean/Spanish buildings and orange groves, it was a delight for him, especially in winter, a sharp contrast to the dark and cold of Berlin. Just before his first trip to Caltech in 1930, the gloomy international outlook had become further clouded by the Wall Street Crash and the onset of the Great Depression. But Einstein initially misread the signs, writing to a colleague 'I believe the present unstable state of affairs in Germany will continue to hold for about ten years. Thereafter it might be good to be in America'.[19]

Detecting his restlessness, the USA coaxed him. During a second winter visit to Caltech in 1931–2, Einstein was approached by Abraham Flexner, the head of a new institute in Princeton. Just before the stock market crash, the Bamberger family had sold its New York department store to Macy's, and used some of the proceeds to found a new institute 'to encourage and support fundamental scholarship'. Flexner was its first director. He asked Einstein if he would be interested in joining the new institute, soon to be called the Institute for Advanced Study, with an annual salary of $15,000, five times the modest sum Einstein had suggested. Meanwhile, Einstein's personal plans accelerated as the Nazi party gained power in Germany. In December 1932, he sailed from Bremerhaven for California. It was supposed to be a three-month trip, but he took thirty pieces of luggage. As he locked his Berlin house, he told his wife that she would not be seeing the place again.[20] His final public appearance in Germany had been a lecture on relativity in Berlin.

In the Spring of 1933, Einstein returned to Europe, shunning a now Nazi-governed Germany and instead renting a house on the Belgian coast, from where he monitored developments in his home country. He did not dare to show up in Berlin. While he had been travelling in the United States, Nazi troopers had broken into his house 'to search for weapons allegedly hidden there by communists',[21] and his bank account had been blocked. Meanwhile he was busy writing letters: he had written to the German authorities to enquire about abandoning the German citizenship which he had acquired automatically after being invited to Berlin in 1914; and he had formally resigned from the Prussian Academy of Science. The Academy replied that it had 'indignantly taken note' of his 'smear campaign' and had 'no reasons for regretting Einstein's departure'. His house on the Belgian coast was guarded. Einstein's sights were now set on taking up Flexner's offer at Princeton, but as he waited in Belgium, other prestigious job offers came in from Jerusalem, Leiden, Oxford, Paris, and Madrid. Weighing their relative attractions was not easy. Writing to Robert Millikan at Caltech, he already had enough experience of America to point out, 'I am a European by instinct and by inclination'. To Paul Langevin in Paris, he wrote of the irony of all his job opportunities at a time when so many of his colleagues were losing theirs.[22]

By this time, the Civil Service Law had been published and the plight of German Jewish intellectuals had been recognized overseas. Hundreds of talented professors, doctors, writers, and musicians were seeking a way to emigrate. On his way from Belgium to the USA, Einstein stopped off in London, where he addressed a rally at the Albert Hall to raise money to support academic refugees from Germany less well known and not as fortunate as himself. Britain too had right-wing extremists, and Einstein was given protection. He arrived in the USA in October 1933, where he was welcomed by Flexner, but public support for the new Nazi regime was growing in a country which had millions of German immigrants, so Flexner warned Einstein to keep his head down and his mouth shut. Einstein duly complied, except on a few famous occasions when he was urged by other refugees not to. His name could still make people listen. In 1940, he formally became a US citizen, the third country to which he claimed allegiance, but enigmatically he retained his Swiss citizenship. With hindsight, Einstein's 1933 move across the Atlantic can be seen as symbolically marking a watershed in international scientific supremacy: the centre of gravity of science moved away from Europe.

The only other contemporary Jewish scientist whose influence was comparable to that of Einstein was Sigmund Freud. Born Sigismund Schlomo (Solomon) Freud into a Jewish family in Pribor, Moravia (then part of Austria-Hungary) in 1856, he qualified as a doctor in Vienna, carrying out research on the sexual anatomy of eels and investigating the analgesic properties of cocaine. As he neared his fiftieth birthday, he turned his attention to the role of the unconscious mind in shaping human personality. While these ideas profoundly affected modern psychology, they were controversial. In 1928, Einstein was asked to support the nomination of Freud for a Nobel Prize for Medicine, but declined, pointing out not only that he found Freud's theories questionable, but also that psychology might not qualify as a suitable topic for the prize.[23] Later, as Einstein began preparations to leave Germany, he was invited by the League of Nations to select anyone he wanted, to discuss with him a problem of his own choosing. He chose as his subject the problem of war, and as his correspondent Freud. Their letters *Warum Krieg? (Why War?)* have been collected and published.[24] In Vienna, Freud had been less visible to Nazis, but his turn would come in 1938, when Nazi Germany annexed Austria.

'Judapest'

The creation of an Austro-Hungarian alliance in 1897 was another of the many major reconfigurations of national frontiers in nineteenth-century Eastern Europe. Just as the newly unified Germany under Bismarck was undergoing a new scientific and industrial revolution, so the new Austria-Hungary was also becoming cultured and prosperous, and its newly affluent twin capitals of Vienna and Budapest attracted youngsters with ambition, particularly Jews from the hinterland of Galicia.[25] By the end of the century, 20 per cent of the population of Budapest was Jewish. In callings such as medicine, law, and finance, the percentage was much higher. Some even spoke of 'Judapest'.[26]

One immediate challenge of moving to Budapest was the Hungarian language. But Jews often had to learn new tongues, and adding another was not a major obstacle. Offsetting any language barrier was Hungary's tradition of Latin culture: the Parliament had debated in Latin until well into the nineteenth century, and the language was taught well in schools.[27] In this fertile breeding ground and in the space of a few years, Budapest produced a remarkable generation of Jewish talent, which itself was pushed into emigration by the political turbulence of the early twentieth century.

This remarkable generation included the aviation pioneer Theodor von Karman (Kármán Tódor in Hungarian, 1881–1983), and Georg von Hevesy (Hevesy György, 1885–1966) who pioneered the use of radioactive tracers, opening up the science of nuclear medicine and who went on to earn the Nobel Chemistry prize in 1943. Michael Polanyi (Polányi Mihály, 1891–1976) first qualified as a doctor before moving to fibre chemistry research in Berlin. Forced to leave Germany in 1933, he moved to Manchester. Later his interests shifted from pure science towards economics and philosophy, and in 1948 Manchester created for him a new chair of social science. (His son John, born in Berlin in 1929 and who arrived in Britain with his parents in 1933, went on to share the 1986 Nobel Prize for Chemistry for his work on chemical dynamics.)

Dennis Gabor (Gàbor Dénes, 1900–1979) was awarded the 1971 Nobel Physics Prize for his invention of holography. Eugene Wigner (Wigner Pál Jenö, 1902–1995) earned that prize in 1963. Journalist and writer Arthur Koestler (Koestler Arthur, 1905–1983) was not a scientist but liked to write about it. Later came cryogenic specialist Nicholas Kurti (Kürti Miklós, 1908–1998), and the eccentric mathematical genius Paul Erdös (Erdös Pál, 1913–1996). There were others. Several even attended the same school, the German-speaking Lutheran Gymnasium. Going to Berlin to study engineering was another popular move. Three of the most significant who took this route were Edward Teller (Teller Ede, 1908–2003), the spiritual father of the H-bomb, the remarkable Leo Szilard (Szilàrd Léo, 1898–1964), and John von Neumann (von Neumann János, 1903–1957), a new paragon of intellect.

Science was not the only talent that Hungary produced at that time: there was also a remarkable generation of film makers—the Korda brothers Alexander (the founder of London Films), Zoltan, and Vincent; Emeric Pressburger (one Oscar and four nominations); Michael Curtiz (five Oscars), and the pioneer war photographer Robert Capa. Adolph Zukor (born Cukor) left Hungary in 1889 at the age of 16 and went on to become founder of Paramount Pictures. The family of William Fox, born Vilmos Fried in 1879, came to the USA in 1880 and provided a new name for the motion picture industry.

Budapest's intellectual and economic boom was brutally halted by the First World War. In 1918, national frontiers were once more rearranged, and a new Hungary, much smaller, was rocked by a series of political upheavals in the vacuum left by the defeated empire. The new 'Red Terror' regime of Béla Kun introduced hard-line communist dogma, transferring ownership of property and commerce to the public domain. Feral gangs looked for rich victims to beat up. In this scrabble for power, communist idealists were overtaken by greedy opportunists. Traditional mechanisms of supply ground to a halt, and with shops no longer operational, city-dwellers had to hike out into the country to purchase from farmers, or simply flee. However the Kun regime was soon replaced by a right-wing rule under Admiral Horthy, former commander of the Austro-Hungarian navy. The Red Terror was replaced by an even more vicious White one. The communist government had been populated by Jews, who were quickly targeted. New lynch mobs roamed the streets, and thousands were killed. It was another warning to leave the country.

Such fast-moving developments led to a general sense of panic and bewilderment, but the young Leo Szilard had an uncanny ability to see what was going to happen, especially when others were confused, and somehow managed be one jump ahead of everyone else. His immediate forebears had progressed from shepherding in the Carpathians, first to farming, then to engineering, the latter step requiring Szilard's German-speaking father to acquire Hungarian. To underline the decision, the family name of Spitz was changed to Szilard in 1900. Even as a teenager, Leo Szilard was able to make better judgements than his parents. In 1914, at the outbreak of the First World War, the Szilard family was returning by train eastwards to Budapest at the same time as troop trains were moving westwards. Szilard's father remarked that the exuberant soldiers seemed to be full of fighting spirit. The young Leo correctly observed that they were full of another kind of spirit, and were simply drunk. Such clarity of judgement remained with him: he attributed it to an ability 'to keep free from emotional involvement'.[28] Later, in the history of nuclear physics, the Atom Bomb, and even beyond, Szilard appears as a sort of intellectual jack-in-the-box who always managed to pop up in the right place at the right time, and tell his sometimes confused colleagues what needed to be done.

Szilard's acute perception saw through the Hungarian fog of hysterical bellicosity and propaganda. Convinced that the Central Powers of Austria-Hungary and Germany would lose the war, he began to make long-term plans. In spite of winning a national mathematics prize in 1916, he abandoned mathematics and physics, convinced that his future lay instead in the booming field of electrical engineering. However these plans were interrupted when he was called up for army service in 1918. His appearance as a cavalry officer at the front was delayed by the epidemic of influenza which swept the world, but Szilard seemed to be a natural survivor. As post-war Hungary was sucked into turmoil, he decided his home country was not the place for studies, and he packed his bags, an operation he would do many more times over the next twenty years. Mobility was Szilard's forte, but this restlessness also hindered his scientific research, which required painstaking perseverance and attention to detail. Flitting from one interest to another like a busy bee in a garden, he quickly sought out the most exotic blooms, pollinating other flowers, but he never stayed put long enough to match the scientific contributions of his more disciplined contemporaries.

Arriving in neighbouring Austria, Szilard sensed that this other remnant of the Austro-Hungarian Empire was not the right place to be either, and in 1920 moved on to Berlin. There, he met other emigrants from Budapest, but far more impressive was the array of scientific talent assembled at Berlin. Dazzled by such a concentration of intellect, Szilard decided to abandon his pragmatic decision to study engineering and revert to physics. As well as being smart, Szilard was precocious. He did not hide in the back row of the lecture room with the other students, and was one of the few who dared to approach the luminaries seated in the front row. In one of Szilard's first contacts with Einstein, he 'suggested' that the master of relativity should talk about new developments in statistical physics. Under the privileged conditions of his appointment, Einstein was under no obligation to teach at all, but he nevertheless took up the suggestion.[29] Szilard's academic precociousness continued: just two years after entering Berlin as an undergraduate, he presented a doctoral thesis, which was accepted. He also exploited Einstein's talent. Szilard could see new technical challenges, and Einstein, best at ideas and mathematical formalism, helped provide the insight to solve them. Remembering his experience at the Swiss Patent Office in Bern, Einstein also understood the need to register these developments. Together, Einstein and Szilard made seventeen patent applications.[30] One of the first was a refrigerator pump which had the advantage of being silent, but which was soon overtaken by other technology. Szilard's vivid imagination had also been fired by the science fiction of H.G.Wells, who had written about the possibility of nuclear bombs in his 1914 book *The World Set Free*. Szilard audaciously travelled to London to enquire if he could handle the European rights for Wells' books.[31]

Inside Germany, Szilard sensed that all was not well politically, and as stock markets throughout the world started to behave erratically, he shifted his savings out of the country. In 1930, he moved onto the first rung of university teaching in Berlin, and obtained German citizenship. Like Einstein, he knew by then that his future lay elsewhere. While Einstein had been a victim of the growing wave of anti-Semitism in Germany, Szilard saw a more global picture. Einstein had been able to pick and choose between handsome job offers: Szilard had to make his own way, first via Britain, but he would meet Einstein again later in the United States. By now an elder statesman of twentieth-century physics, Einstein became a spectator in the rapid development of nuclear physics. On the other hand, Szilard could see nuclear implications just as clearly as political ones. But he knew that few people would listen to him, even if he carried an important message. So he resolved to use the highly conspicuous Einstein as a political chesspiece. Their common objectives went on to be much more important than refrigerator motors. Szilard's respect for Einstein was reciprocated: in 1933, Einstein had written a letter of recommendation: Szilard, in his view, was 'one of those men, rich in ideas, who create intellectual and spiritual life wherever they are'.[32] A master manipulator, Szilard could be both charming and disdainful. When a well-wisher tried to contact him in 1942, saying, 'we have had no news of you for a long time, but I am sure that you are deeply engaged in the war effort', he replied ten months later, 'I am answering your letter a little prematurely, as it is not yet one year old'.[33] A tribute described him as 'a restless, homeless spirit. He owned no property and few books. Hotel lobbies, cafés, Jewish delicatessens, poor restaurants, and city pavements were the settings for his discussions'.[34]

NOTES

1. Sturtevant, A.H., *A History of Genetics*, NY, Cold Spring Harbour Press, 2001.
2. Pais ELH, p.114.
3. Pais SL, p.37.
4. Pais SL, p.521.
5. AIP history interview, 4568 P. Debye.
6. Pais SL, p.185.
7. Gilbert, p.3.
8. Pais ELH, p.166.
9. Gilbert, p.400.
10. Pais SL, p.186.
11. Pais SL, p.240.
12. Szanton, p.70.
13. Frisch, p.34.
14. Pais ELH, p.254, citing the *New York Times.*
15. Kamran, M., *Einstein and Germany*, Sang-i-Meel, Lahore, 2009, quoting Clarke, R., *Einstein, His Life and Times*, NY, Random House, 1995.
16. Clark, R., *Einstein, His Life and Times*, NY, Random House, 1995, p.393.
17. Pais ELH, p.158.
18. Pais ELH, p.185.

19. Pais ELH, p.187.
20. Pais ELH, p.190.
21. Pais ELH, p.191.
22. Holton, G., The Migration of Physicists to the US. In *The Muses Flee Adolf Hitler*, Jackman, J.C. and Borden, C.M. (eds), Smithsonian Institution Press, Washington DC, 1983.
23. Pais SL, p.514.
24. Einstein A. Freud S., *Warum Krieg?* Diogenes, Zurich, 1972.
25. A region now covered by Poland and the Ukraine.
26. Macrae, p. 41.
27. Hargittai, p.14.
28. Hargittai, p.8.
29. Hargittai, p.44.
30. Pais SL, p.489.
31. Rhodes MAB, p.14.
32. Clark, R.W., *Einstein, His Life and Times,* NY, Random House, 1995.
33. SPSL archives, Box 342.
34. Shils, E., *Leo Szilard—A Memoir*, Encounter, New York, 1964.

5

The fall of German science

Science can be influenced by the same foibles of predilection and fashion as are literature, art, and music, but ultimately its objectivity cannot be a matter of taste, or be decided by a lay jury. There are vivid examples of scientist martyrs: in the seventeenth century, Galileo had been warned by Rome to stop preaching that the Earth moves around the Sun, and was confined to house arrest for the rest of his life. His contemporary Giordano Bruno was not so fortunate, and was burned at the stake.

Although he did not suffer so much, Albert Einstein was another such victim. His strong influence could make people move in different directions. Just as unlike electric charges attract and like ones repel, so Einstein's insight could inspire some, while his unconventionality antagonized others. He was a major influence on a new generation of quantum scientists, whose collective efforts went on to show that the atom works in unexpected ways. These cannot be expressed in words at all. Atomic quantum mechanics lies hidden behind a heavy mathematical curtain. Its results shrink from the light of everyday experience, throwing up instead apparent conundrums and contradictions.

However this nonconformity further exasperated others, already irritated by the discord between Einsteinian relativity and deep-rooted 'commonsense' notions. The uneasiness affected students too. A student of the new quantum physics, Victor Weisskopf, later to make important scientific contributions and become a key member of the US atomic bomb project, had initial misgivings. He wrote: 'I felt that [quantum mechanics] was an esoteric theory far removed from everyday experience. I considered giving up abstract science to do something more important'.[1]

In Germany, a reactionary scientific movement resisted this new unorthodoxy. It was led by two pitiful figures, Johannes Stark (1874–1957) and Philipp Lenard (1862–1947). Both had Nobel prizes, but to their chagrin, the advances which had merited these awards had been overtaken and overshadowed by subsequent developments, particularly those of Einstein. This poisoned their outlook. Stark had discovered in 1913 how the characteristic fingerprint of atomic spectral lines is shifted when an electrical field is switched on. This 'Stark Effect' earned him the 1919 Nobel Prize for Physics, one year after Planck, and reinforced Germany's international scientific prestige. It also provided a fresh example of the interplay between electrons—the carriers of electricity—and Einstein's light quanta. In 1907, Stark had invited Einstein to contribute an encyclopaedia article on relativity, and was, briefly, an enthusiastic supporter of Einstein's revolutionary quantum ideas.[2]

But as quantum physics moved on, Stark became increasingly sceptical of its growing incomprehensibility. This clashed with the 'before your very eyes' vividness of his own experiments. For him, science had clearly taken a wrong turning, and had become blocked in a Jewish-influenced impasse. There was also some personal humiliation: Stark was piqued when the public acclamation following Einstein's Nobel Prize eclipsed anything that he had experienced three years before. As well as nursing these grudges, Stark did not prosper. His post-Nobel career had stalled: he had invested his prize money in a porcelain factory, and his enthusiasm for dubious projects and his growing resistance to progressive physics led to him losing his professorship at Würzburg. Unable to find another university post, and with the depression years ruining his porcelain business, he seethed. Ignored and humiliated, he became convinced that he was another victim of the purported Jewish conspiracy which had humbled the proud German nation, and which was now polluting science with outlandish relativity and quantum mumbo-jumbo. Einstein and others mocked him as 'Giovanni Fortissimo'.[3] His 1922 book *The Contemporary Crisis in German Physics* reinforced the rift between racial and objective science.[4] 1914 Nobel laureate Max von Laue said of it, 'All in all, we would have wished that the book had remained unwritten'.[5]

In the cultural cleansing following the Nazi accession to power in 1933, science too had to be made to conform. Concepts that were incomprehensible and enigmatic were bad enough, but with so many of their advocates in addition being Jewish, unacceptable scientific modernism was lumped together with Judaism. Quantum mechanics and Einsteinian relativity were dismissed as abominable *Judenphysik* which had no place in a culturally pure Germany.[6] Scientific objectivity was sacrificed on the altar of misguided ethnic idealism. How could Einstein and the others claim to explain what we see around us, when their underlying ideas were incomprehensible? Under the Nazis, Stark was transformed from a scientific Cinderella into the Prince Charming of a new 'Aryan science', purged of Jewish pollution. His voice grew steadily more audible, and as the Nazi drums beat louder he became in turn President of the *Physikalisch Technische Reichsanstalt* in Berlin and the *Notgemeinschaft* emergency funding agency. Influential in the Education Ministry, he had a voice in deciding which research projects went forward, and promoting new schemes, especially in the military sector.

Stark's fellow malcontent was Philipp Lenard, twelve years his senior. Born in Bratislava, he was in fact Hungarian. In spite of having worked with Heinrich Hertz, who had Jewish ancestry, and editing his collected works, Lenard scorned Jewish scientists. He had first made his mark on science in the late nineteenth century, before the quantum idea had even been introduced by Planck. A precursor to the atomic physics age had been cathode rays, the eerie glow produced by a high voltage in a vacuum. Modern textbooks say that these rays—in fact a stream of electrons—were discovered by J.J. Thomson in England in 1897, but there is a case for them to be viewed as a German discovery.[7] Even though Lenard went on to earn the 1905 Nobel Prize for his work, one year before Thomson, a perceived British 'media scoop' of cathode rays and electrons rankled some Germans.

Meanwhile Lenard had moved on from electrons to look at their implications for atomic structure.[8] He then studied the photoelectric effect, discovering totally unexpected results,[9] which were eventually explained by an upstart named Albert Einstein. This explanation seemed to generate more scientific interest than Lenard's discovery. Again he saw one of his major contributions to science snatched from under his nose, and his resentment grew. This intellectual humiliation could have been underlined by personal indignity. For a time, Lenard was deformed, his head clamped to his shoulder, as a result of incorrect medical treatment for a childhood affliction, requiring in turn painful surgery.[10]

Lenard's bitterness exploded when Einstein received the 1921 Nobel Prize for his explanation of the very effect which Lenard had discovered. In spite of the fact that he already had a Nobel Prize, Lenard felt he had been snubbed, relegated to a scientific has-been, another victim of the stab-in-the-back conspiracy. To add to the misery, his son died of malnutrition and his investments in national bonds became worthless.[11] After the assassination of Rathenau in 1922, the German government ordered a national day of mourning. When Lenard refused to close his Heidelberg laboratory and to lower the national flag, irate students demanded that he be thrown in the river. He was also temporarily suspended from duty. Furious, frustrated, and embittered by his own downfall and the increasing emphasis on seemingly unintelligible theories, he joined the nascent Nazi movement and helped launch the Aryan physics campaign. Its ultra-conservatism was enshrined in the four volumes of his racist *Deutsche Physik*. With no mention of quantum mechanics and relativity, its foreword said that it was aimed at 'the seekers of truth'. His lectures, illustrated with impressive demonstration experiments, became popular with students, and Lenard further groomed his image by wearing a huge swastika on his lapel.[12] At the inauguration of the Philipp Lenard Institute in Heidelberg in 1936, some German scientists and politicians started a campaign that succeeded in banning the eminent British science journal *Nature* from libraries. One year later, Minister of Science Bernhard Rust made the ban official.[13] After the Second World War, Lenard at 83 was deemed too old for punishment.

In May 1933, Stark was appointed as the first Nazi-approved President of the *Physikalische Technische Reichsanstalt*. Describing the event as a 'Great day for science', Lenard lamented in the *Völkscher Beobachter*,[14] the 'darkness' that had beset physics, with the very foundation of science—the observation of Nature—having been devalued and fallen into oblivion. This misfortune was the direct result of a 'massive penetration' of Jews into positions of scientific authority. Einstein was the only name cited. However his 'ideas', together with those of other Jews, had become misguided 'theories' which had set science on the wrong road. Lenard reproached his fellow scientists for having allowed these 'Relativity Jews' to settle in Germany, and for considering them as 'good Germans'. After rejoicing that the culprits were leaving the universities, and even the country, after the Civil Service Law, Lenard concluded his vicious diatribe with an attack on the Nobel prizes, which in recent times had become 'increasingly of contestable value'. However Stark's flat-footed enthusiasm eventually antagonized even the Nazis, who found him too disruptive, and he was eventually sidelined.

His final humiliation came after the Second World War, when in 1947, he was condemned to four years' hard labour, with German scientists giving evidence against him.[15]

Another exponent of *Deutsche Physik* was Bruno Thüring (1905–89), who, as an astronomer at Munich, and later Vienna, was particularly critical of Einsteinian relativity, claiming that there were parallels between relativity and the Talmud.[16] Other examples were Lenard's student Rudolf Tomaschek (1895–1966), and Wilhelm Müller (1880–1968), later to succeed Arnold Sommerfeld at Munich. While Lenard and Stark were promoting their cause, Ludwig Bieberbach (1886–1982) did likewise for '*Deutsche Mathematik*': unorthodox developments such as Cantor's set theory were shunned, and the intuitive was preferred to the esoteric.

(It is one thing to falsify arguments, or accuse others of doing so, but quite another to falsify actual results. The whole of science is based in observation. Results can be wrong for many reasons, but Emil Rupp (1898–1979) actually falsified them. Although an apparently gifted student of Lenard, he chose to work in industry rather than embark on a university career. In 1926 he claimed to have shown the first evidence for the wave–particle duality required by the new quantum mechanics. Einstein initially welcomed these results, but other scientists were unable to duplicate them and became suspicious. Undeterred, Rupp went on to make even bolder claims, which were soon rejected. After having been considered one of the leading lights of German physics for a decade, Rupp and his work were disgraced.)

The Nazi practice of cherry-picking science became more difficult when it came to medicine. As one scornful commentator put it: 'A Nazi who has venereal disease must not allow himself to be cured by Salvarsan, as it is the invention of the Jew Ehrlich. He must not even take steps to find out whether he has this ugly disease, because the Wasserman reaction which is used for the purpose is the discovery of a Jew [Albert Neisser]. If he has toothache he will not use cocaine, or he will be benefiting from the work of a Jew, Carl Koller. Typhoid must not be treated, or he will have to benefit from the discoveries of Jews'.[17]

The 'White Jew'

It was both an irony and a tragedy that the bigoted trap set by Stark and Lenard ensnared a German patriot and one of the nation's finest scientific minds, almost destroying him. Werner Heisenberg, born in Würzburg in 1901, became for a time as much a victim of Nazi oppression as his Jewish colleagues, a Jew by association rather than blood.

His father, a respected teacher of Greek at Munich University, volunteered for service on the Western front in 1914, despite being over 40. His son, too young for military service, joined his school's *Wehrkraftverein* cadet force and was assigned first to farm work. Afterwards, Werner Heisenberg joined the *Neupfadfinder* (New Pathfinders) a version of the Boy Scouts, and in some respects a precursor of the later Nazi youth movement. Overwhelmed by the political and moral confusion in Germany, the Pathfinders tried to return to more fundamental values, such as a love of nature and an appreciation of culture, together with unquestioned allegiance to an acknowledged leader, usually someone with war experience. Heisenberg's involvement with the movement was no passing affair. In 1919, his group assisted loyal troops from Berlin sent in to quell Bavarian rebels.

Although an enthusiastic and conscientious student with a gift for mathematics, Heisenberg was clumsy with laboratory apparatus and his first physics degree in Munich was almost a disaster. Hesitantly, he turned to theoretical research in the mathematical stratosphere of Göttingen. Here, his objective was to explain the complex patterns of spectral lines—the fingerprints of effervescent atoms—using Niels Bohr's system of quantum jumps. Heisenberg's mathematical atoms became increasingly complicated, wheels within wheels, reminiscent of a Ptolemaic astrolabe. Frustrated, he stripped away the atomic complexity and looked at the case of a simple quantum oscillator. How did a single electric charge emit quanta of radiation?

In his mathematical workshop, Heisenberg devised a set of rules to simulate atomic vibrations. Heisenberg's professor, Max Born, later recognized these rules as being those for the manipulation of mathematical constructs known as matrices.[18] Heisenberg had never encountered matrices before, and unknowingly had worked out their mathematics for himself. But what did his mathematical manipulations represent? In the summer of 1925 came an epiphany on the blustery North Sea island of Helgoland. He realized that his matrix oscillator was simulating the emission or absorption of energy as well as radiation, and that its mathematics reproduced Bohr's mysterious quantum jumps.

His scheme worked, but nobody could explain why. Many were sceptical. The mystery soon deepened when Erwin Schrödinger produced an alternative picture based on a wave equation. Schrödinger's explanation was easier to comprehend, a way of stuffing particulate waves into the available atomic space. But an additional puzzle was how they both worked: one could subscribe to Schrödinger's 'wave mechanics' or to Heisenberg's 'matrix mechanics' and arrive at the same result by two seemingly different routes.

As Heisenberg continued to explore his scheme, its enigma deepened. His mathematical pointers were supposed to indicate where an oscillator was and what it was doing. But instead of pointing, they slid about unpredictably. If one was fixed, another immediately came unstuck. It was the 'Uncertainty Principle' which now bears Heisenberg's name, a badge of modern science. Such paradoxes were the trademark of the new quantum mechanics. Just 26 years of age, Heisenberg moved to a professorship at Leipzig.

Stark and Lenard noticed. Heisenberg's matrix mechanics was just as unintelligible and unwelcome as Einstein's relativity, and his work and his name quickly became targets for the rabid bombast of *Deutsche Physik*. Heisenberg was 'the spirit of Einstein's spirit',[19] and was branded a scientific heretic. There had to be a Jewish link. Nazi authorities looked back over five generations of Heisenbergs (originally Heissenberg) to search for signs of Semitic ancestry. Such genealogical exactitude was normally reserved for candidates to the crack SS (*Schutz Staffel*) security corps. No links were found, but Heisenberg's meteoric career nevertheless stalled.

Just as David Hilbert in Göttingen towered over German mathematics, so did Arnold Sommerfeld (1868–1951) in physics. Schooled in the mathematical powerhouse of Göttingen, Sommerfeld ensured that German physics developed in step with its mathematics. In 1906, he had been headhunted by Wilhelm Röntgen for a new professorship in Munich. Röntgen chose well. Sommerfeld's school went on to nurture several generations of eminent German scientists, including six Nobel laureates (one was Heisenberg). Sommerfeld himself helped launch the quantum picture of the atom, and was frequently nominated for a Nobel Prize, but was never successful. His achievements were so numerous and wide-ranging that it was difficult to edit them into a single Nobel Prize citation, and after a time his achievements became overshadowed by those of his students.

In the early 1930s, Sommerfeld was approaching retirement. He himself wanted Heisenberg as his successor in Munich. The 1933 Civil Service decree had left a swathe of empty seats. One was the vacancy created by Einstein in Berlin, where Schrödinger had also left in disgust: another was caused by Max Born's sudden departure from Göttingen. Heisenberg was a natural candidate for any one of these prestigious posts, but the university authorities no longer had a free hand. Candidatures now had to be rubber-stamped by the Ministry of Education, and Heisenberg was initially passed over. Frustrated, and angered by the forced departure of so many people he admired, he considered leaving Leipzig in protest. Planck persuaded him to stay. It was unpatriotic, he admonished the young professor, for such talent to resign merely in protest.[20] Planck told Heisenberg that Hitler had assured him that 'nothing would be done beyond the new Civil Service Law that will impede our science'.[21]

Comforted, Heisenberg looked forward to being Sommerfeld's successor in the city in which he had grown up. He also attempted to further the cause of his science in the Nazi press, but quickly discovered that this only made his own position more perilous. He watched in dismay as German science increasingly became a shambles. Posts left empty after the 1933 departures remained unfilled. After the Civil Service Law, while Sommerfeld and a few others did their best to shelter Jewish colleagues, many scientists preferred to sign a declaration of support for Hitler.[22] Rudolf Hess, Hitler's deputy, established a League of university teachers as a Nazi route into the academic community and to better monitor 'progress'.[23]

Heisenberg was one of the signatories of a 1936 letter urging the government to fill the still-vacant posts. National science was stagnating, they warned. While it was natural for a few posts to be vacant at any one, time, in the mid-1930s some 17 out of a hundred professorial posts were unfilled.[24] For a time, it looked as though Berlin had reneged, with initial signs of approval for Heisenberg's move to Munich. But these moves immediately revived the full fury of the demonic Stark, whose personal ambitions had meanwhile switched from porcelain to gold; convinced that there was truth behind ancient myths that the precious metal lay hidden under the hunting grounds of Teutonic gods on central German heathland.

On an exploratory trip to Munich in 1937, Heisenberg was shocked to read a vituperative full-page article by Stark in *Das Schwarze Korps*

(The Black Corps), the official newspaper of the elite SS. In a grotesque parody, it highlighted the role of 'White Jews' in science, those who were not Jews by blood, but effectively became so by adopting Jewish ideas and culture. It condemned Heisenberg's 'dictatorship of the grey theory'. After pointing to real Jews, Einstein and Haber, Stark then accused their sympathizers, Sommerfeld and Planck, of preferring to appoint Jews to university positions rather than Germans: Heisenberg's group in Leipzig was an example of such corruption. For good measure, Heisenberg was bracketed with the dissident Carl von Ossietzky.

However, a single newspaper article can still be overlooked if it does not fall on fertile ground. To strengthen their case, Heisenberg's opponents called on Heinrich Himmler, head of the powerful *Gestapo* (*Geheime Staatspolizei*—Secret Police). With his personal honour now at stake, Heisenberg also wrote to Himmler. But instead of using normal channels, where such a letter would probably be intercepted before it reached its intended destination, Heisenberg sent it via his grandfather, a personal friend of Himmler's father. It was a wise move. Meanwhile Heisenberg kept his options open by contacting an envoy specially sent to Germany to ask whether he would be interested in a job at Columbia University, New York.

But the Nazis had an even more malevolent trick up their sleeve. The idea was to accuse Heisenberg of homosexuality. In those days, such perceived 'criminals' were being rounded up and put into internment camps. One Nazi ploy was to accuse victims of something worse, and offer escape if they pleaded guilty to a lesser 'crime'. The plan was to exonerate Heisenberg of being homosexual if he admitted to peddling 'Jewish physics'. However Himmler intervened personally, acknowledging that the case had been referred to him by his father:[25] Heisenberg was deemed 'decent', and was allowed to continue to teach physics as he thought fit—provided that he did not mention the names of any Jews. There could have been a much higher price to pay.

As though to hinder his scientific work further, Heisenberg was mobilized. As German troops marched into the Sudetenland, the German-speaking territory that post-1918 had become part of Czechoslovakia, the German army was placed on maximum alert. However the European climate of appeasement defused the volatile situation, and Heisenberg returned to Leipzig. Soon afterwards, Wilhelm Müller, an exponent of *Deutsche Physik*, finally stepped into Sommerfeld's now-cold shoes in Munich. Sommerfeld described his successor as a 'complete idiot'.[26] The profanation of one of the nation's most prestigious posts epitomized the descent of German science.

In 1939, the redeemed Heisenberg travelled to the United States.[27] Knowing how he had been treated, his American hosts were puzzled, even dismayed, by his refusal of their tempting job offers. However, after all the political meddling, and despite his shabby treatment, Heisenberg felt that he still had a responsibility towards German science. It had not been an easy decision: he said that he 'almost envied' his colleagues who were forced to leave.[28] But he and his wife had bought a new home in Bavaria, and he wanted to enjoy it. A month after he left New York, Germany was at war again, and Heisenberg was drawn into other developments. After his treatment at the hands of Stark, he had been overshadowed in the scientific establishment by Adolf Butenandt, whose ascent to power was boosted by his 1939 Nobel Chemistry Prize for work on sex hormones.

During the ensuing war, Heisenberg was to play a key role in the German quest to investigate nuclear fission. In one famous attempt to break the wartime isolation, in 1941, Heisenberg travelled to Copenhagen to meet his former colleague Niels Bohr. The idea was for Heisenberg to glean whatever he could about the weapons potential of nuclear fission. There has been much speculation about this meeting,[29] and its significance has perhaps been over-emphasized. In Copenhagen, Heisenberg was in occupied territory and there could be no open exchange of views. His friendship with Bohr became another of victim of the war.[30] The Nazi era transformed Heisenberg completely. Victor Weisskopf says that on meeting Heisenberg after the war 'he was completely transformed. ... Before the war he had always struck me as innocent, ... youthful and enthusiastic. ... But when I saw him again, even his complexion had changed.' Weisskopf compared Heisenberg's metamorphosis to that of Dorian Grey in the novel by Oscar Wilde. While Dorian Grey was destroyed by his own abandonment to hedonism, Heisenberg's fate was the result of his own idealism and patriotism.

Tree in the wind

Even after Einstein became a world figure, Max Planck remained Germany's senior scientific statesman. As the nation emerged from the trauma of the First World War, Planck set out to ensure that it would retain its scientific supremacy. This was helped by the award of the 1918 Nobel Prizes for Physics and for Chemistry to Planck and Fritz Haber, respectively. The pair also helped establish the *Notgemeinschaft der Deutschen Wissenschaft* (Emergency Committee for German Science) to unify and coordinate a scientific effort divided into many specialist sectors, and in danger of fragmenting even more amid the post-war upheavals. Both had suffered in the war years. Haber had been tainted by his poison-gas war effort, and his wife had committed suicide. Planck's wife had already died in 1909, his son was killed at Verdun, one of his twin daughters died after childbirth, and her sister died after marrying her brother-in-law. (More misery was in store for him in the next World War.)

Like the biblical Job, Planck somehow climbed above such personal hardships. Embarrassed and frustrated by the continual post-war ills which befell his country, he broadly agreed with the stated objective of the Nazis—to restore the nation to its former greatness. But he did not concur with the way they wanted to do it. As Hitler came to power and the viciousness of Nazi prejudice became apparent, Planck was shocked. But he felt that the national cause was too important to be sacrificed to moral principles. Already 74 when the Nazis came to power, he had shed most of his formal responsibilities, but was still regarded by his scientific colleagues, and by the political establishment, as a figurehead. Others urged him to help Jewish scientists, but after talking with Hitler, Planck realized that efforts to get the *Führer* to change his mind were hopeless. Whatever their feelings about Nazis, many saw how their own careers would benefit by the departure of so many talented Jewish scientists.

Planck advocated a 'tree in the wind' approach, bending with the applied force rather than resisting it.[31] At a major meeting in May 1933, he declared, 'Everyone who truly loves our dear Fatherland and who not only lives for the moment but also thinks about the future, will immediately understand that today everybody ... must stand ready for action, and that today there can be only one motto for all Germans, a motto which [Hitler] himself has solemnly announced'.[32] But behind the scenes, Planck continued to help individual Jewish scientists as best he could.

Some highly principled scientists emigrated even if they did not have to.[33] However, Planck urged those inclined to follow to reconsider: if they went, they would fill positions abroad which would otherwise be open to those who were forced to leave. One who did emigrate was

Erwin Schrödinger, who had replaced Planck when he retired as professor of theoretical physics at Berlin in 1927. Discovery in science is often the prerogative of the young, but Schrödinger, like Planck, was exceptional, in many ways. He was born in 1887 in Vienna into a family centred around a semi-aristocratic mother. His maternal grandmother was English: as a child he spent time in England, and could speak English even before he mastered German.[34] He had no Jewish ancestry.

At the turn of the century, Vienna, like Budapest, was a booming, cosmopolitan city. In art, a strong avant-garde movement was headed by Gustav Klimt, and the city attracted many aspirant painters. Among the less successful was the young Adolf Hitler, who tried to hawk his paintings in the street after being rejected by the Vienna Academy of Fine Arts. Viennese science was dominated by the monumental figure of Ludwig Boltzmann (1844–1906). With atoms still only a theoretical idea, he nevertheless showed how the statistics of such invisible constituents could explain the behaviour of vast assemblies of them. But Boltzmann's genius hid a tortured inner self, and he committed suicide. His successor, Fritz Hasenöhrl (1874–1915) was called to active service at the outbreak of the First World War, despite being 40 and a distinguished university professor. Eager to see action, he pressed to be sent to the front, where Austrian forces were facing the Italian army. In July 1915 he was killed by a grenade.

After receiving his degree in 1910, Schrödinger underwent a year of military service assigned to a unit of siege artillery. Returning to Vienna as a teaching assistant, he had begun to make his mark in the research field when war was declared. Like Hasenöhrl, Schrödinger was sent to the Italian front. For several years he remained near Görz (now Gorizia in Italy), shelling Italian positions. Half a million soldiers were wiped out in this carnage. In 1917, Schrödinger, promoted and mentioned in despatches, was sent back to Vienna to teach meteorology to a new kind of artillery unit, anti-aircraft gunners. In 1920, he married Annemarie Bertel, whom he had met ten years previously. She was intellectually dazzled by her husband, and they remained an odd but devoted couple, despite Schrödinger's increasingly unconventional sex life. Annemarie (Anny) compared living with her husband to living with a racehorse.[35]

Searching for a successor to Hasenöhrl, Vienna ignored Schrödinger, who began to look instead towards German universities. In the early

1920s, he shuffled through a series of posts in Jena, Stuttgart, and Breslau (Wroclaw) before an unexpected offer came from Zurich. It was prestigious—the post which Einstein had held before leaving for Berlin in 1912. In addition, neutral Switzerland was isolated from the post-war upheaval. In 1925, 38 years old, Schrödinger felt he had reached his career plateau, and was confident enough to step back from science and write *Meine Weltansicht (My World-View)*. With his new status, his lifestyle acquired four important but very different facets: science, culture, vacations, and sex, all of which he took very seriously.

During the war years, the theory of the atom had made a major advance when Niels Bohr and Arnold Sommerfeld had taken the quantum ideas of Planck and Einstein and applied them to atoms. Electrons inside atoms, they said, appeared to live on some kind of staircase of energy, emitting or absorbing light quanta when they took a 'quantum jump' downwards or upwards. Schrödinger, like Heisenberg, joined the throng of eager researchers looking to extend these ideas. The particle-like behaviour of radiation, which Einstein had introduced in 1905, had yet to find any complementary mirror image. If light—radiation—could look like particles, then could the reverse be true—could particles also act like waves? The pointer came in the 1924 doctoral thesis of a then-unknown French student, Louis de Broglie, who had spent the war years operating radio equipment in the Eiffel Tower in Paris. De Broglie[36] wrote down an equation suggesting how atomic particles, such as electrons, could look like waves. Before his final examination, one of de Broglie's examiners sent a copy of the student's work to Einstein, who said that it 'lifted a corner of the great veil'. In October 1925, Schrödinger gave a talk in Zurich on de Broglie's new idea.

During the following Christmas vacation, Schrödinger left his wife, who remained more or less patiently in Zurich, and went up into the Swiss mountains with 'an old girlfriend from Vienna'.[37] When he was not otherwise engaged, a newly inspired Schrödinger wrote down the equation which now bears his name and which describes the wave behaviour of a particle, rather than merely suggesting it. In his mountain love-nest, Schrödinger grappled with the implications of his equation, with its mysterious 'wave-function', a new property of particles, but which itself was not corpuscular.[38]

Putting aside the mysterious wave-function, the next problem was to solve the new equation. It threw up obscure algebraic forms which had been studied by mathematicians a hundred years before and which lay buried deep inside textbooks. Back in Zurich, Schrödinger did his homework and showed how to understand the behaviour of the hydrogen atom. His career now went into a fresh orbit, and in 1927 he moved to Planck's chair in Berlin. There, he and Einstein, of a similar generation, and both expatriates in the German capital, soon became close friends, sailing on the lakes in Berlin's countryside, two great minds afloat.

Berlin in the 1920s was a perfect place for a hedonist like Schrödinger. He was obsessed with women, and they noticed. He was strikingly unconventional: he did not wear a Germanic suit, collar, and tie to give lectures. A 1933 photograph (Plate 3) of physics Nobel prizewinners taken at Stockholm railway station shows the other laureates, Paul Dirac and Werner Heisenberg, in staid dark coats, clutching their hats. On the right is the flamboyant figure of Schrödinger, in plus-fours, a fur-trimmed coat, and his bow tie awry. He appears to be dressed for some fancy-dress party. While the other two Nobel scientists, much younger, are accompanied by their mothers, Schrödinger is alongside his wife Anny, who by that time had ceased to play a conventional matrimonial role in his life.

In 1925, Schrödinger was giving private mathematics classes to the fourteen-year-old non-identical twins Itha and Roswitha Junger, who had failed their school mathematics examinations. It seems odd for a distinguished professor to volunteer for such a mundane job, but Schrödinger had his reasons. Of the twins, Itha was the more attractive, and Schrödinger was soon seeing her alone, and for more than just mathematics. Even after he moved to Berlin, his infatuation with the now-sultry Itha blossomed. 'Sex relations in all their splendour,' he enthused, 'and with all the beauty of the world that follows from them'.[39]

He had considered divorce, but Anny did not think such convention was necessary. In Zurich she had already responded by embarking on an affair with Hermann Weyl, an influential German mathematician who had worked closely with Einstein. Not wishing to be left out, Mrs Weyl went to bed with the Swiss atomic scientist Paul Scherrer. Itha regularly came to Berlin, where Schrödinger introduced her to Planck and to Einstein. Itha was the first woman to become pregnant by Schrödinger, who wanted children, but she had an abortion to avoid the inevitable complications. She could see that Schrödinger was incapable of fidelity. He now wanted Hilde, the wife of Innsbruck university scientist Arthur March.

Schrödinger's career was interrupted in 1933 with the rise to power of the Nazis. Already saddened by the departure of his friend Einstein, he was horrified by the new wave of anti-Semitism. For the 1 April boycott of Jewish businesses, one target was the Jewish-owned Wertheim department store, the first of its kind in Berlin, which had been condemned in some quarters as a 'bazaar'. Its large windows were smashed by stones, and the store was ordered to close, although uncomprehending shoppers still turned up. Anyone thought to be Jewish was a target for Nazi troopers. Schrödinger was walking past as one hapless shopper became a victim, and he boldly intervened. The troopers' attention was instantly diverted. The Austrian professor would have been beaten up too, had it not been for the fact that one of the troopers was a Berlin physics student who knew who he was.[40]

The Civil Service Law had alerted headhunters from Britain and the United States, an opportunity to recruit young talent. One of these headhunters was Frederick Lindemann, later Viscount Cherwell and Winston Churchill's scientific advisor in the Second World War. Lindemann had studied under Nernst, had excellent contacts in Berlin, and was able to summon funding from Imperial Chemical Industries. On his list of people to see was Schrödinger, whose promising young research assistant, Fritz London, now had no future in Germany and who Lindemann wanted to recruit for Oxford.

Fritz London (1900–54) and his brother Heinz (1907–70) came from a wealthy assimilated Jewish family. Both made their mark in the quantum theory of molecular forces, and went on to suggest that quantum effects were not confined to individual atoms and molecules. They could also extend across certain materials in bulk, explaining new phenomena (superconductivity), seen only at low temperatures. Forced to leave Berlin in 1933, Fritz moved initially to Oxford with Lindemann, then to Bristol and Paris, before finding in 1939 a career position at Duke University, North Carolina. Heinz also moved to Oxford in 1933. From there, he transferred to Bristol, where he was rounded up and briefly interned as an 'enemy alien', in 1940 (see Chapter 7).

However, when Schrödinger learned that Fritz London had not yet made up his mind whether to take up Lindemann's offer, he proposed going to Oxford himself. Lindemann was astonished, but reluctant to let such an unexpected catch off the hook. As an afterthought,

Schrödinger asked Lindemann if a position could also be found for Arthur March. Schrödinger wanted to have all his comforts. While Schrödinger was ready to sacrifice his own career because of Nazi anti-Semitism—which did not concern him directly—he was absurdly dishonourable in his personal affairs, both a conscientious objector and a shameless libertine.

When the new position was finalized, Schrödinger left Berlin without handing in any formal notice to the university. The only clue was a brief message to the departmental office to say that he would no longer be giving lectures. It was a slap in the face for the Nazis. He left Berlin in his new car, with Anny driving, taking a roundabout route through Europe, stopping in the Tyrol where the Marches had a summer residence. Hilde and Schrödinger disappeared on a cycling tour, from which she returned pregnant. In November 1933, the Schrödinger/March contingent arrived in Oxford, where Hilde's daughter, Ruth, was born six months later. Schrödinger's arrival in Oxford also coincided with the news that he had won the 1933 Nobel Prize for Physics.

There were few women at Oxford in those days. If women appeared, as for the Magdalen College Christmas festivities, they were relegated to the viewing gallery, as though at a synagogue or a mosque. Schrödinger scorned these misogynist ways and openly went around with Hilde, while still living with his wife. Schrödinger had not advertized this unconventional arrangement before arriving at Oxford, but bachelor Lindemann was astonished, as well as being angry, to discover that he had been duped into employing March simply to enable Schrödinger to continue his sex life. The short-term grant which supported March was not renewed, and the Marches, including baby Ruth, returned to Austria.

With Hilde no longer available, Schrödinger soon set up with Hansi Bauer-Bohm, from a wealthy Vienna family, who was working as a photographer in London. Hansi and Schrödinger were not strangers—he got to know her during another mountain holiday, in fact while Hansi had been on her honeymoon. Anny obligingly moved to London so that her husband could have more time alone in Oxford with Hansi. But he was now concerned about his academic future: Oxford at the time was not as prestigious a science centre as Berlin or Cambridge.[41] His hurried departure from Berlin meant that he could not return to Germany, and he had upset Oxford with his bizarre lifestyle, which clouded even his Nobel Prize reputation. Later in 1933 he turned down the offer of a post at Princeton, despite the fact that he would be near Einstein, with whom he was examining new quantum minutiae. This collaboration had uncovered startling new mysteries, such as the now legendary 'Schrödinger's cat' enigma. But Schrödinger wanted to remain in Europe. An attempt to transfer to Edinburgh failed after Schrödinger went to inspect the University wearing one of his Alpine fancy-dress outfits, and the post went instead to a Jewish refugee, Max Born, later to earn the 1954 Nobel Prize for physics. Schrödinger's request for a residence permit went astray, while that of Born took a more direct route.[42] In 1936 Schrödinger returned to Austria, this time to Graz, the capital of the province of Styria. In the fifteen years since he left the country, it had changed, mirroring the developments in neighbouring Germany.

On return, one of Schrödinger's first moves was to install Hilde March and their daughter in Graz. Anny, renewing her own family and other ties, could now spend a lot of time away. The following summer Schrödinger also taught at Vienna. In 1938 came Germany's uninvited annexation of Austria, when Hitler returned to his homeland in triumph. Immediately all Jews in Austrian university posts were subject to the same laws as those of Germany. Although Schrödinger was not a Jew, the Nazis had not forgotten his snub in Berlin. He was 'advised' to make a public declaration of his support for the regime. In his 'Acclamation of the *Führer*' which appeared in the press on 30 March 1938, Schrödinger grovelled 'amid the exultant joy which has penetrated our country'. This declaration brought him the worst of both worlds. Schrödinger's apparent obsequiousness brought criticism from abroad, but on the other hand was not enough to appease the Nazis. Deprived of the opportunity to do so in Berlin in 1933, they sacked him in Austria in 1938.

With his incoming mail now being intercepted, Schrödinger and his still-faithful wife fled Graz with just a suitcase, taking a train to Rome, where the Italian scientist Enrico Fermi introduced Schrödinger to the Vatican. From its Papal Academy, he requested help from Eamon de Valera, the Prime Minister of Ireland and President of the Council of the League of Nations in Geneva. On his way to meet de Valera, his train was held up before reaching the Swiss border and the Schrödinger's belongings were searched. They carried almost no cash (although in their haste to sack him, the Nazis had overlooked Schrödinger's Nobel

Prize money). With de Valera's assistance, Schrödinger took up a post at the new Institute for Advanced Studies being established in Dublin in neutral Ireland. Schrödinger duly moved in, with his wife, Hilde March, and daughter Ruth all in tow. There, the ageing scientist widened his scientific horizon and wrote *What is Life?*, based on his popular lectures at Dublin which examined how physics and chemistry could account for the underlying processes of living organisms. The book was to have a major impact on a new generation of molecular biologists. In Dublin, the indefatigable hedonist also fathered two children with Irish mothers. In 1956, a symbolic figure, he returned to Vienna, where he died in 1961. His grim-faced image appeared on Austrian 1000 schilling banknotes, together with the symbol Ψ, the wave-function introduced by his equation.

No J for Pauli

While some scientists derided quantum mechanics as *Judenphysik*, others, more tolerant, preferred to call it '*Knabenphysik*' (boy-physics). Schrödinger, 40, was too old for a generation of scientists still in their twenties. As well as Heisenberg, another prominent *Knabenphysiker* was Wolfgang Pauli (1900–58). Pauli's father's family in Vienna originally carried the name Pascheles, and were Sephardic Jews from Spain who had first settled in Prague. Pauli's father formally converted to Catholicism and had his son baptized as a Catholic.[43] The young Pauli grew up with little idea of his family roots.[44] To cloud the issue further, in 1911 Pauli's father left the Catholic Church and embraced evangelism. During his university career in Germany, Pauli was at one time Max Born's assistant in Göttingen, and later discovered the 'Pauli Exclusion Principle' which explains how atomic electron structure underlies the pattern of chemical elements seen in the Periodic Table.

In 1928, Pauli became Professor at the Swiss Federal Technology Institute in Zurich. When Germany annexed Austria in 1938, the Austrian Pauli technically became a German national. Although Switzerland was supposedly neutral, he still felt nervous. His attempt to obtain Swiss citizenship was turned down, and to go anywhere he had to obtain a German passport. Here, he ran up against the race laws, and had to produce evidence of his family background. Pauli was nevertheless relieved when the German consulate in Zurich counted him as half-Aryan rather than half Jewish, and awarded him a passport without the huge 'J' stamp which marked the travel documents of Jews.[45] Pauli travelled by a roundabout route to the safety of the Institute for Advanced Study in Princeton. Once there, he took no part in wartime nuclear science. After the award of his Nobel Prize in 1945, Pauli could have stayed in the USA, but preferred to return to Zurich, where he eventually earned the Swiss nationality which he had long sought.

Before his wartime departure from Zurich, Pauli had employed a distinguished dynasty of assistants, many of whom went on to become famous in their own right. One was Nicholas Kemmer (1911–98), born in St. Petersburg, who also learned quantum physics at Göttingen under Born. With his career blocked in Germany, Kemmer moved to Britain in 1936, working initially at London's Imperial College. He later became involved with wartime nuclear research, first in Cambridge, then in Canada. When Born retired and moved back to Germany in 1953, Kemmer inherited his Edinburgh professorial chair. Other Pauli assistants who emigrated in the 1930s were Victor Weisskopf, Rudolph Peierls, Felix Bloch, and Hendrik Casimir, all of whom will appear later in this book.

Less well known, but nevertheless illustrative, is the case of Valentin Bargmann, who worked with Pauli in Zurich in the mid-1930s. He studied in Berlin, but had to leave in 1933, transferring to Zurich. Not able to get a post there, he moved to live with his parents, who had moved to Lithuania in 1933. There he was able to procure a US visa through a family connection,[46] and moved to Princeton, where he worked first with Einstein, and after the outbreak of war, with John von Neumann. They worked on gas dynamics (shock waves). Bargmann understood fully the implications. Small tailor-made explosions were needed to detonate the power of a fission bomb. After the war, he collaborated with Eugene Wigner to develop a new equation for subnuclear particles.

NOTES

1. Weisskopf, p.31.
2. Beyerchen, p.103.
3. *Stark* means 'strong' in German.
4. Cornwell, p.108.

5. Cornwell, p.109.
6. For an excellent description, see Beyerchen.
7. Pais IB, p.131.
8. Pais IB, p.182.
9. Pais IB, p.380.
10. Beyerchen, p.82.
11. Cornwell, p.106.
12. AIP history interview, W. Elsasser 4590.
13. Hossfeld, U. and Olson, L., 'Freedom of the mind got Nature banned by the Nazis', correspondence in *Nature*, **443**, 2006.
14. Berlin edition, 15 May 1933.
15. Cornwell, p.416.
16. *Forschungen zur Judenfrage*, **4**, 1941, 134–162.
17. cited in Roth JCC, p.215.
18. An example of a matrix is a grid showing the exchange rates between different currencies, with rows and columns labelled 'dollar', 'euro', 'yen', 'pound', etc. The numbers are the 'matrix elements'.
19. Cassidy, p.351.
20. Medawar and Pyke, p.156.
21. Eckert, M., Werner Heisenberg, Controversial Scientist, *Physics World*, Vol 30, November 2001.
22. Noakes and Pridham, p.250.
23. Noakes and Pridham, p.250.
24. Hentschel, K., *Physics and National Socialism, An Anthology of Primary Sources*, Birkhauser, Berlin, 1996. Cited in *Peer Effects in Science*, Waldinger, F., University of Warwick paper, 2010, http://www2.warwick.ac.uk/fac/soc/economics/staff/academic/waldinger/research/
25. Beyerchen, p.163.
26. Eckert, M., *Physics World*, November 2001.
27. The first time having been in 1929.
28. Beyerchen, p. 63, taken from Heisenberg, *Der Teil und das Ganze*, Piper, Munich, 1969, p.209.
29. The play, *Copenhagen*, by Michael Freyn.
30. Weisskopf, p.130.
31. Heim, Sachse, and Walker, p.3.
32. Max Planck Gesellschaft archives, cited in Heim, Sachse, and Walker, p.27.
33. for a list, see Moore, p.272.
34. Moore, p.14.
35. Moore, p.131.
36. pronounced 'de Broy'.
37. Moore, p.194.

38. It was later interpreted as the square root of the probability of finding the particle.
39. Moore, p.255.
40. Interview with Frau Schrödinger, cited in Moore, p.265.
41. Interview with Frau Schrödinger, AIP oral history transcript 4865.
42. Moore, p.319.
43. Ernst Mach, his godfather, presented a baptismal cup.
44. AIP oral history, P. Ewald 4596.
45. Enz, C., *Wolfgang Pauli, Wissenschaftlicher Briefwechsel,* Vol 3, U. Chicago Press, 1993.
46. http://www.princeton.edu/~mudd/finding_aids/mathoral/pmc02.htm

6

Plagues

When the Israelites were slaves in Ancient Egypt, Moses implored Pharaoh to 'let my people go'. The refusal prompted a series of plagues: blood, frogs, lice, beasts, cattle disease, boils, hail, darkness, and locusts. With Pharaoh still incompliant, came a tenth plague, the slaying of firstborn Egyptian children. This made him change his mind, and the biblical Exodus went ahead. In Nazi Germany, the respective roles were reversed. The Law for the Restoration of the Civil Service heralded a very different exodus. In the Biblical one, the enslaved Israelites wanted to improve their lot by emigrating, and a charismatic leader, Moses, engineered their departure. In contrast, Hitler, a twentieth-century Pharaoh, wanted to get rid of a contented Jewish population which did not want to go. The Civil Service Law was the first of a series of Nazi plagues aimed at convincing reluctant Jews to pack their bags and leave. Unlike the mass emigration of the biblical Exodus, that from Middle Europe in the 1930s was disorganized and fragmented, with no equivalent of Moses and with individuals managing as best they could. Some, like Einstein, were triumphantly successful. Others were not. Those who could not or would not join that exodus succumbed to the final plague, the Holocaust.

For most people, the Civil Service Law was like being woken up by being doused with a bucket of ice-cold water. Before the Nazis came to power, German Jews, naturally hypersensitive about their status, had noted deep currents of anti-Semitism at work. While this was unwelcome, it was not new, and most of them nevertheless had felt secure and comfortable enough to dismiss it as a nuisance. Now the 'non-Aryan' reference in the Civil Service Law and its derivatives made things clear. Persecution of Jews had become official. If anybody had any doubt about what was still just a piece of paper, it had been enthusiastically underlined by the almost simultaneous *Judenboykott*. Helped by such demonstrations, and by years of propaganda, the new legislation fell on pockets of fertile ground. Many students began boycotting lectures by

Jewish teachers. Lists of allegedly Jewish teachers and students began to appear on the walls of public toilets. Nazi students picketed lecture halls. Led by Education Minister Bernhard Rust, speeches and press articles claimed that 'Science for a Jew does not mean ... creative organization, but a way of destroying the culture of the host people.... Positions were vacated for them to allow them to pursue their parasitic activities, which were then rewarded with Nobel Prizes'.[1] Nobel laureates were becoming targets.

The Civil Service Law and collateral legislation affected thousands of professionals. Confused and disoriented, they learned that the new Nazi-speak could mean that they no longer gave lectures, or even have official contact with their former colleagues or students. Sometimes they could not even use the university library. Instead of a salary, they received a pension, never more than 75 per cent of their salary, but often much less, particularly if they had previously earned money through lecturing fees. More junior staff, the *Privatdozenten*, who had to rely totally on lecturing fees, had no tenure and received no pension at all. They were simply sacked.[2] Most of the victims—'helpless and disbelieving'[3]—looked on, hoping that the wave of calamity was a bad dream and would not last. Some reacted immediately, not even waiting for a personal letter to arrive. Others resisted. Thousands were involved, and each would have a story to tell.

The biochemist Max Rudolf Lemberg, born in Breslau in 1896, had volunteered for war service before embarking on his studies. After a period in industry, he became a lecturer at Heidelberg. Under the 1933 ruling, as a teacher who earned lecturing fees he was dismissed from his assistantship and 'placed on leave', despite his war record. However he was reluctant to abandon experiments still underway in the laboratory, and was instructed to do the absolute minimum so that 'valuable material would not be lost'. He left Heidelberg for a year in Cambridge, leaving a message to the Heidelberg authorities: 'I have to tell you that abroad I will always remain loyal to Germany, and will retain the same love for my country and my countrymen that made me fight for Germany in the war.' He later emigrated to Australia, where he became a founder member of the Australian Academy of Science.[4] Another victim at Heidelberg was Leonardo Olschki, a noted authority on medieval and renaissance manuscripts. Born in Verona, Italy, in 1885, he moved to Germany and took up a position at Heidelberg in 1913. In the winter of 1932, he was on sabbatical leave in Rome, where, after the publication of the 1933 laws he was told to stay until further notice. Replying that he could find no record of any such new legislation at the German embassy in Rome, his dismissal had been initially rescinded. However a subsequent letter informed him that he was formally to be placed on leave in December 1933, but the letter was sent to his old address in Heidelberg, and returned unread. The leave notice was promptly brought forward to November. He left Rome for the USA.[5]

Cumbersome machinery of official racial classification began to turn. With official records now open for inspection, there was compliant snooping and finger-pointing. Newspapers began to publish lists. Some snared by the new legislation solicited friends in high places. Some were warned in advance. One university teacher had an old friend, a former army officer, who was working in the secret police, compiling lists of undesirables. With his loyalties divided, he alerted his old comrade that it was time to make a dignified exit while there was still time.[6]

The trickle of German Jews who followed Einstein's example and left the country suddenly increased, if not to a flood, then certainly to a steady stream. The *Hilfsverein der deutschen Juden* (Organization for Assistance to German Jews) in Berlin reported in the first days of April 1933 that their staff of eleven could no longer cope. However, from April to December of that year nearly 6000 applicants were given assistance to emigrate to European destinations, while complementary schemes for assistance abroad were still being coaxed into action. Some of these early applicants were themselves immigrants to Germany from countries further east, especially Poland, who had lost their original nationality.[7] The profile of the applicants also changed. Once it had been those simply seeking work. In the queue now were professionals: lawyers and notaries, engineers, doctors, officials, and academics from all disciplines. The situation worsened in 1935, with the introduction of the Nuremberg Laws. Anyone who could not muster a clean *Ariernachweis* was in danger. A few public-spirited objectors who tried to complain were themselves dismissed.[8]

There were initial loopholes which granted exemption to those with war service, and to those whose fathers or sons had died in the war. Seniority also counted: those who had been in service since 1914 were also initially exempt. Younger people had no such place to hide, and even those benefiting from exemptions could find themselves cold-shouldered or in deeper trouble. The prominent Jewish scientist Carl Neuberg (1877–1956) had made important advances in the study of fermentation enzymes, and was appointed Director of the Kaiser Wilhelm Institute for Biochemistry in 1925. His war medals made him formally exempt from the Civil Service Law, and he made no attempt to conceal his origins. But sacking a technician who had been harassing other Jewish employees did not advance his career. Neuberg was threatened by Nazi troopers and vilified by the authorities. After asking Max Planck for help, he finally stepped down and emigrated to the USA in 1937, but judged that he had made the move too late, and had lost his place in the rush.[9]

However the initial exemptions to the Civil Service law were soon annulled. They had at first been dismissed by Hitler as a detail, but the Nazis were surprised to discover how many prominent professionals and other key administrators had war medals at home. From 1935, even before the Nuremberg Laws, dismissals were judged on a case-by-case basis.[10] Those already retired were also affected. Richard Willstätter (1872–1942) had been born into a Jewish family in Karlsruhe. After studying in Munich, he moved to a post in Zurich, where he elaborated the chemical structure of plant pigments, particularly chlorophyll, which earned him the 1915 Nobel Prize for Chemistry. In 1912, he became Director of the new Kaiser Wilhelm Institute for Chemistry in Berlin, and in 1916 moved to a senior professorial chair in Munich. There, he was close to the fomenting Nazi movement and saw the growth of its rabid anti-Semitism. In 1924, at the age of 53, he became its target, and to the consternation of other colleagues and students, he retired. When the Nazis came to power ten years later, they came looking for him, and he emigrated to tranquil Switzerland.

Before leaving, Willstätter was visited by Selig Hecht (1892–1947), Professor of Biophysics at Columbia University, New York, whose parents had emigrated to the USA from Galicia around the turn of the century. Although educated in the USA, Hecht's sensitivity to the European mindset acquired through his upbringing had been underlined by research experience in England, Sweden, and Italy.[11] After taking up his Columbia position, he went on another extended European trip in 1933, where he saw at first hand the initial repercussions of the Civil Service Law. While shocked to see such a drastic move, Hecht was also surprised by the 'callousness of the population', who moreover dared not talk about it among themselves for fear of being incarcerated in a concentration camp. According to Hecht, Willstätter was not surprised by the 1933 legislation, which merely underlined what many had been saying for years, usually in private, but sometimes publicly. Hecht also met the distinguished Jewish chemist Kasimir Fajans, born in Warsaw in 1887, and a one-time colleague of Willstätter. In 1935, Fajans moved to the USA, taking up a post at Michigan.

Otto Meyerhof (1884–1951) was from a Jewish family in Hanover, and in 1922, shared the Nobel Prize for Medicine with the British scientist A.V. Hill for his work on muscle metabolism. He became Director of the Kaiser Wilhelm Institute for Medical Research. As this privately funded institute was initially not affected by the Civil Service Laws, Meyerhof retained Jewish staff and students, and tried to shelter them when the law was later extended. Eventually this was 'brought to the attention' of the authorities, and in 1938 he fled, first to Switzerland, and then to France, where he was interned after the German invasion, but managed to travel through Spain to the United States. One of his research collaborators was Fritz Lipmann (1899–1975), born into a Jewish family in Königsberg, and who moved to the USA in 1939, becoming Professor of Biological Chemistry at Harvard Medical School. In 1953, he was awarded (with Hans Krebs) the Nobel Prize for Medicine for his work on the function of living cells.

Georg Schlesinger (1874–1949) was a celebrated mechanical engineer at the Technische Hochschule, Berlin, Charlottenburg, and during the First World War had an influential position in the Spandau armaments factory. In 1933 he was targeted as a Jew and imprisoned on charges of espionage and treason, but was well treated and allowed to continue writing a book. Released from prison, he formally retired and emigrated to Switzerland. In 1939 he moved to the UK, teaching at Loughborough College of Technology. He died in London.

Biochemistry and molecular biology are outside the scope of this book, but several influential quantum physicists, notably Leo Szilard and Max Delbrück, later moved from nuclear physics to play important roles in these fields. Among others who made this research shift was Konrad Bloch, born in Neisse, Germany, in 1912 and who left Germany for Switzerland in 1934 while still a student, later moving to the USA, becoming Professor of Biochemistry at Harvard, and earning the Nobel Prize for Medicine in 1964 for his work on cholesterol and fatty acids. Bernard Katz, born into a Jewish family in Leipzig in 1911, left Germany in 1935 after graduation, moving to the UK to work with A.V. Hill.

He went on to earn the 1970 Nobel Prize for Medicine for his work on nerve synapses. Ernst Chain, born into a Russian-Jewish family in Berlin in 1906, fled anti-Semitic Germany in 1934, moving to the UK. He shared the 1945 Nobel Prize for Medicine for his key role in discovering the therapeutic properties of penicillin. His story is a saga in its own right.[12]

A Nobel Chemistry Prize was awarded to Gerhard Herzberg in 1971 for his work on the spectra of molecules. Born in Hamburg in 1904, he lost his post at Darmstadt's Technische Hochschule because his wife was Jewish. He explained his predicament of having been dismissed for marrying a Jewess before 'the revolution', and stated 'nobody who has committed such 'race treason' can get new employment'.[13] With Carnegie Foundation money organized through the Zurich office of the Emergency Committee for German Scientists Abroad (*Notgemeinschaft Deutscher Wissenschaftler in Ausland*), he left Germany in 1935, eventually moving to Canada after spending a few years in Chicago.

Along with such Nobel triumphs, there were also tragedies. The accomplished mathematician Leon Lichtenstein, born in Warsaw in 1878, came to Berlin in 1894 as a student, and obtained a university post in 1910. He founded mathematical journals and moved to a professorial chair in Leipzig in 1922. As a Jew, he lost this post in 1933, but had no need to move. Already ill, he died of heart failure on 21 August. More dramatic was the case of Paul Ehrenfest (1880–1933), born in Vienna to a Jewish family that had moved from rural Moravia.[14] He first studied chemistry, but was steered towards physics by the influence of Ludwig Boltzmann. In 1900 he moved to Göttingen, where he was further inspired by David Hilbert and Felix Klein. There he also met his future wife, a talented Russian physicist. As she was a Russian Orthodox Christian, their marriage in Vienna meant that Ehrenfest formally had to revoke the religion of his birth. The couple lived in St Petersburg for several years, and played an important role in continuing the work of Boltzmann after he committed suicide in 1906. Making strenuous efforts to get a university post in Germany, Ehrenfest met Einstein in Prague, where the two men quickly bonded. Both were accomplished violinists as well as scientists. In 1912, Einstein recommended Ehrenfest as his successor in Prague, but the authorities did not appreciate his religious ambivalence. Instead, he moved to Leiden, to succeed the relativity pioneer Hendrick Lorentz. Thereafter, Einstein was a regular visitor to the Dutch city. After finding his feet academically, Ehrenfest's life began to fall apart. In 1918, a son was born with Down syndrome; his wife began to spend a lot of time in Russia, and physics moved into unfamiliar territory. At the same time, Ehrenfests's friends and colleagues in Germany, notably Einstein, became the victims of Nazi anti-Semitism. When this reached a new climax in 1933, and after unsuccessfully trying to shoot his disabled son, Ehrenfest committed suicide.

Felix Auerbach, at Jena, had once been a colleague of Erwin Schrödinger. Born into a Jewish family in Breslau in 1856, Auerbach became a keen promoter of art, and lived in a house designed by Walter Gropius, where he continued to live after his retirement in 1927. In March 1933 he and his wife committed suicide: they 'Left their earthly existence full of joy'.[15] Arnold Berliner, an industrial physicist and close friend of Paul Ehrlich and Walter Rathenau, had, in 1912, persuaded Berlin publishers Springer to set up the new journal *Die Naturwissenschaften*. As a Jew, Berliner lost his job in August 1935, but already 72, preferred to stay put in Berlin. Paul Rosbaud was his successor. In 1942, he committed suicide when the Nazis came to take him to the gas chambers.

Felix Hausdorff (1868–1942), the son of a Breslau merchant, moved with his family to Leipzig, where he went on to become a distinguished mathematician. He made important contributions to the study of topology and, after Georg Cantor, helped develop set theory. Under the pseudonym of Paul Mongré, he also wrote some notable philosophical literature. Already 65 when the Civil Service Law was introduced, he was initially exempted because he had been in his position before 1914. When this loophole was blocked, he tried to emigrate to the USA. Having missed the initial wave of emigration, and with his age an obstacle, he did not succeed. Under the threat of 'deportation' to a concentration camp, he and his wife took their own lives in January 1942. Wilhelm Traube (1866–1942) was a gifted chemist who synthesized several substances, notably caffeine. Of Jewish origin, he proclaimed himself to be an evangelist. However, neither this nor his scientific achievements could prevent him from losing his emeritus Berlin professorship in 1935, age 69. In September 1942, destined for a concentration camp, he too committed suicide.

Nobel Prizes and suicides were extreme cases. The reactions to the Civil Service Laws were as diverse as the laws themselves and their application. Some victims were told outright that they had to leave, some saw their names published in newspapers, others knew that their cases had been placed on hold, pending a decision. Sometimes their right to teach was revoked as an interim measure. This lack of transparency was also a ploy to induce people to leave of their own accord. One victim of being deprived of teaching status was Lothar Nordheim at Göttingen, originally born in Munich in 1899. Although describing himself as 'Lutheran', he had Jewish blood and moved to the Netherlands in 1934. He later went to the USA, becoming Director of a Division of the Oak Ridge Laboratory, which produced fissionable material for the Atomic Bomb.

The scope of the initial 1933 Civil Service Law was extended by a swathe of subsequent amendments and additional legislation. In 1938 it was extended geographically as well. In that year, German troops marched into Austria. The country had made its own way politically for more than a century, but its inhabitants spoke German and even the name *Österreich* ('Eastern Empire') tacitly invited *Anschluss* (annexation). The rest of the world looked on disapprovingly but did nothing. Effectively overnight, Austrians became Germans, subject to the same laws. Austrian Jews who had been working in German universities and who had closed their eyes to the new developments became *persona non grata*. Jewish emigration from Austria quickly became highly organized. For the Nazis, Adolf Eichmann took over this infrastructure, and its success soon led to his techniques being adopted in Germany itself.[16] However, with so many Jews wanting to get out of Austria, Switzerland, the nearest haven, was soon overwhelmed. Alarmed, the Swiss authorities tried to impose some kind of restriction, but any such moves would have to be reciprocally imposed on Swiss citizens travelling to and from Germany. Technically, the old Austrian passports were no longer valid: Austrians travelling abroad needed to have German passports, and German documents were subject to the race laws. The outcome was that after 1938, the German passports of Jews were stamped with a large J.[17]

One Austrian-Jewish Nobel Laureate victim was Otto Loewi (1873–1961), born in Frankfurt, but who moved in 1904 to Austria to continue his research career, taking up Austrian nationality. His work on the chemistry of muscle function earned him the 1936 Prize for Medicine. The Nazis were unimpressed by these achievements and ordered him to leave, after ensuring that his Nobel Prize money stayed behind. He went to the United States, via Brussels and Oxford, and did not reappear in Austria until 1958. Also ensnared by the *Anschluss* was Victor Hess (1883–1964), who in 1912 had made a series of daring high-altitude balloon ascents which found that there were more stray subatomic particles 5000 metres above the Earth than at sea level. This showed that such subatomic debris came from outer space, rather than being radiated outwards from the Earth as had been assumed. Hess's 'cosmic rays' earned him the 1936 Nobel Prize for Physics. However he did not approve of Nazis and, in addition, his wife was non-Aryan. He was imprisoned for nine days and his Nobel Prize money confiscated. Erwin Schrödinger, by this time in Oxford, explained how the Nazis in Austria handled such cases: 'the authorities take mild decisions at first (for example letting a victim keep his pension rights), thus allowing the news to spread, and then making more severe decisions (for example cancelling all pensions rights)'. There was a move to invite Hess to the UK, but with financial support by that time stretched thin, it was feared that Hess would become stranded there.[18] He left Innsbruck for Fordham University, New York, in 1938.

Other notable Austrians who lost their positions included Karel Przibram of Vienna's Radium Research Institute, and Hans Thirring of Vienna University, both former colleagues of Erwin Schrödinger. Przibram spent the war years in Brussels and later returned to Vienna. His zoologist brother Hans was less fortunate and died in the Theresienstadt concentration camp. Thirring, a non-Jew, was targeted because of his pacifist leanings and evident admiration for the work of Einstein and Freud. After wartime industrial work, he returned to the University of Vienna and later became involved in politics. Responding to Red Cross enquiries about missing persons in 1946, he added 'we are always hungry'.[19] Leo Hayek, born in Prague in 1887, was the director of the sound-recording archive of the Viennese Academy of Science. In 1939, he applied to the central emigration bureau in Vienna. Described in his references as a 'most ingenious mechanic', Hayek moved to the USA, where he set up as a consultant and repairer of scientific instruments, but soon got a regular job in the speech clinic of Ohio University.[20]

Bruno Touschek, born in Vienna in 1921, was the child of an Austrian officer and a Jewish mother, who died when he was ten, but he kept in close contact with her family in Italy. Although a talented student, Touscheck's university career was wrecked by *Anschluss* and then the outbreak of war. Drawn into wartime work in Germany on electronics and prototype particle accelerators, along similar lines to those of Richard Gans,[21] he was eventually arrested by the Gestapo on 17 March 1945. After he and other prisoners were force-marched to Kiel, he fell at the side of the road, was shot at, and left for dead. After the war, he went on to play a major role in the development of particle accelerators in Europe.[22]

The influential writer Stefan Zweig, born in Vienna in 1881, the son of a wealthy textile merchant, left Austria before its annexation. The most visible Jewish refugee from annexed Austria was Sigmund Freud. His work had already fed the flames of the 1933 book-burnings. He said, 'What progress we are making. In the Middle Ages they would have burned me. Now they are content with burning my books'.[23] The Gestapo repeatedly searched his house, and his daughter was taken away for questioning. After confiscation of his property and payment of the hefty emigration tax, Freud took the train to London, but died the following year. Many of his siblings, less well known and less fortunate, perished in the Holocaust. However his architect son Ernst had left Germany for the UK in 1933, taking his young sons Lucien, later an artist, and Clement, who became a colourful politician, celebrity chef, writer and broadcaster.

One Austrian intellectual colossus had left long before. The philosopher Ludwig Wittgenstein was born in 1889 into a Viennese family: his father controlled the iron and steel industry of the Austro-Hungarian Empire, and was a major patron of Viennese art and music. This highly active but authoritarian environment deeply affected the Wittgenstein children. Three of Ludwig's brothers committed suicide, while his brother Paul became a famous concert pianist. His career should have ended when he lost an arm during the First World War, but he went on to become even more famous as a one-handed player.

Ludwig's Wittgenstein's early aim, to study physics in Vienna, dissolved when Boltzmann died. Like several other young Austrians with scientific ambition, he turned instead to engineering, going to study first in Berlin, then in Manchester, where his work in aeronautics introduced him to the study of mathematics under the tutelage of Bertrand Russell in Cambridge. Shocked by the outbreak of the First World War, Wittgenstein volunteered to fight for Austria-Hungary, seeing action on several fronts. Winning medals for bravery, he was finally taken prisoner, where he began work on his epic *Tractatus*, an exploration of the relationship between language and reality.

After working as a humble schoolteacher and a gardener in Austria, he returned to Cambridge in 1929, this time on the academic payroll. With no other qualifications, his *Tractatus* earned him an entry ticket.

With the Nazi annexation of Austria came the rapid implementation of the Nuremberg Laws and a general scramble to acquire the requisite racial clearance papers. The complacent Wittgensteins were perturbed to discover a Jewish family background and shocked to realize what this meant. They had considered themselves staunch Christians and had readily joined in contemporary anti-Semitic banter. To get off the hook, they transferred massive amounts of their assets to the German *Reichsbank*, their final clearance mysteriously being signed by Hitler himself in the days leading up to the German invasion of Poland in 1939.[24] During the ensuing war, Ludwig Wittgenstein left Cambridge to work as an obscure hospital assistant. The lifelong iconoclast died in 1951.

However, Jews were not the only Nazi targets. Other victims included active socialists, communists, pacifists, and dissidents of all kinds. Many of them initially fared worse than apolitical Jews. Instead of merely being fired from their jobs, they were arrested and imprisoned in the new concentration camps. One was Carl von Ossietzky, a newspaper editor and secretary of the German Peace Society. He was awarded the Nobel Peace Prize in 1935, despite Nazi attempts to block the decision. Under arrest, he was unable to travel to Oslo to receive the award, an act which the Nazis threatened would lose him his nationality. Nazi Minister Hermann Göring offered von Ossietzky his freedom if he refused the Prize, but von Ossietzky would not do so. Lying ill in hospital, he issued a statement: 'After much consideration, I have made the decision to accept the Nobel Peace Prize. I cannot share the view put forward by the *Gestapo* that in doing so I exclude myself from German society. The Nobel Peace Prize is not a sign of an internal political struggle, but of understanding between peoples. As a recipient of the prize, I will do my best to encourage this understanding and as a German I will always bear in mind Germany's justifiable interests in Europe'. Still in custody, he died of tuberculosis in 1938. After the von Ossietzky episode, Hitler decreed that Germans could not receive any Nobel Prize, but the Stockholm authorities chose to ignore the ruling. Germans were good at science: Gerhard Domagk was selected for the 1939 Prize in Physiology and Medicine, Richard Kuhn for the 1938 Chemistry Prize, and Adolf Butenandt for the 1939 Chemistry Prize. After much confusion, and some convoluted manoeuvres, they later received their diplomas and medals after the Second World War.

Nazi politics liked to use carefully stage-managed milestones—the Reichstag fire, the book-burnings, the Nuremberg rallies. Another, in November 1938, presaged a further plague on helpless Jews. Herschel Grynszpan, a Jew of Polish origin and living in Germany, had fled to Paris. There, the validity of his complicated papers lapsed, and he was ordered to return to Poland, a country he did not know. With no valid travel documents and knowing that his family, expelled from Germany, was similarly stranded on the Polish border, he strode into the German Embassy in Paris and shot an official. He chose to do it on a notable anniversary in the Nazi calendar: the fifteenth anniversary of the beer-hall *putsch,* which had propelled the Nazis onto the political scene in Munich. Grynszpan's crime triggered a furious backlash. It was one of the few examples of Jewish direct action against Nazis. (Another had been in 1936, when Wilhelm Gustloff, a prominent Nazi working in Switzerland, was shot by the son of a rabbi, but Nazi retribution was muffled that year during the Olympic Games in Berlin.) In the ensuing *Kristallnacht*, organized mobs smashed the windows of Jewish shops, and torched synagogues. Ninety Jews were killed and some 20,000 rounded up and taken to concentration camps. Petrified by *Kristallnacht* and the new outburst of hate, and with passports stamped with a large 'J' now only valid for leaving the country, the exodus accelerated.

One victim of *Kristallnacht* was a Jewish student of David Hilbert, the talented Max Dehn, whose situation as Professor at Frankfurt seemed to have escaped the attention of the Nazis. Preferring to stay in Germany, he was arrested but subsequently released due to prison overcrowding. As a refugee, he preferred the proximity of Norway to the traditional pilgrimage to the United States, but when Norway was invaded by German troops in 1940, Dehn deemed an Atlantic crossing too dangerous and travelled to the USA via the Trans-Siberian railway to Vladivostok, and then by boat to Japan and California, a route also chosen the same year by the eminent Austrian mathematician Kurt Gödel, who went to join Einstein and John von Neumann at Princeton's Institute for Advanced Study.

The contrasting fate of two notable brothers underlines the radical change in the status of German Jews in those years. The career of Gershon Scholem, born in Berlin in 1897, reflects the cultural vigour of the prosperous Jewish community in the early twentieth century.

Instead of continuing the trend towards assimilation, against his father's wishes, Scholem turned to rabbinical studies, and later the mystic rites of *Kabbalah*. His parents had called him Gerhard, preferring a German name to a traditional Hebrew one as a signal of their intention for him to integrate into society, but the renegade student changed it to display his intention to do otherwise. As well as religion, Scholem studied mathematics, philosophy, and Hebrew in Berlin and Munich: the fact that he was able to follow this academic path displays the extent to which Jewish culture had become integrated. (During his studies, he met the philosopher Martin Buber (1878–1965), who had made a diametrically opposite choice, moving from an orthodox Jewish family background in Vienna to secular studies.) Scholem left Germany in 1923 to join the new Hebrew University of Jerusalem, where he founded its department of Jewish mysticism. In stark contrast, Scholem's brother Werner was a hard-line communist who became a member of the *Reichstag* in 1924. As an obstreperous Marxist as well as a Jew, Werner Scholem was among the first to be taken into 'protective custody' by the Nazis in 1933. He was shot in Buchenwald concentration camp.

Göttingen

The most visible target for the Nazi attack on academe was the great university of Göttingen. Traditionally, its students had immense respect for their teachers. Once its physics students had greeted a visiting professor by assembling at his house at dusk and chanting Planck's new radiation law.[25] But in 1933 its intellectual foundations were under siege. Nazi troopers marched through the town, throwing stones at windows and beating up victims.[26] On 26 April, the *Göttinger Tageblatt* newspaper announced that six university professors were being placed on leave, among them Max Born, Richard Courant, and Emmy Noether.[27] Eventually more than half of its mathematics and science faculty were to go,[28] many of them intellectual giants. The mathematician Richard Courant (1888–1972) claimed exemption because of his war service—he had been shot and gassed at Verdun, such was the level of cannon-fodder used in that war—but his politics were alleged to be too left-wing. He decided to resign, ironically after he had been told that he was formally exempt from the Civil Service Law.[29] After a year at Cambridge, he moved on to New York University, where he founded an institute for graduate studies. In 1964, the institute was renamed in his honour.

The talent of Max Born (1882–1970) had been spotted at Göttingen by David Hilbert, who wanted him as his assistant. Born came from a cultured and prosperous Jewish family, whose history mirrors well the assimilation of German Jews. His paternal grandfather was brought up in a Polish ghetto under Prussian, rather than Russian, control. Here, boys initially had to learn at the Jewish *cheder* ('room' in Hebrew, in Yiddish a religious elementary school), but able students were soon able to progress to public schools. Census-takers periodically combed the country for signs of assimilation. Could a person speak German? Had they assumed a family name? Born's grandfather fared well, and progressed to study medicine in Berlin.[30] Born's father, Gustav, was an influential physiologist at the University of Breslau. In Born's childhood, assimilation appears to have progressed to a point where religion played no part in family life. Without really knowing what it meant, he was taught the Lord's prayer by his sister's wet-nurse.[31] With his marriage in 1913 to a woman of Jewish descent but whose family had become enthusiastically Lutheran, Born encountered religious dilemmas. After several compromises, he declared himself a strict Lutheran. Pulled in opposing directions by personal feelings and the need to present himself to the world, he later explained, 'In the end I made up my mind that a rational being ... ought to regard religion as a matter of no importance. ... I did not want to live in a Jewish world, and one cannot live in a Christian world as an outsider. However I made up my mind never to conceal my Jewish origins'.[32] For him, to say that he was German and Jewish was as natural as other Germans saying that they were Catholic or Protestant. However, when Nazi anti-Semitism emerged, Born, along with many other assimilated Jews, had to reappraise his situation. Talking of his family, he said, 'they are really Jews for today's German Laws, but they have never really thought about it before. They are missing, and I also, every single emotional connection to Judaism'.[33] For the Nazis, being Jewish was not a matter of personal choice.

A major career boost for Born was an invitation to work with Minkowski in Göttingen, but its impact evaporated when Minkowski died of appendicitis in 1909. Born delivered a eulogy at the funeral. In line with his earlier decision not to conceal his roots, Born's formal habilitation application before becoming a teacher at Göttingen began,

'I, Max Born, of the Jewish religion...'[34] After a period of military service, Born worked in Berlin, alongside Einstein. They were only a few years apart in age. As well as their science, Born enjoyed playing piano to Einstein's violin. He moved back to Göttingen in 1921, where his wide-ranging scientific interests turned to the new quantum theory, and his students, among them Werner Heisenberg and Wolfgang Pauli, were to bring adolescent quantum mechanics to a new maturity. Less well known than Einstein, Born was not a public target, but on the other hand he knew what grass-roots Germans were thinking. His fragile health meant that he spent a lot of time in hospital, where he overheard conversations about 'the problem caused by the Jews and the promise that Hitler offered Germany'.[35] This was later underlined when he was woken at home in the night by telephone calls shouting '*Juden raus*' (Out with the Jews) and '*Juda verrecke*' (May Judah perish).[36]

Göttingen's reputation meant that professorial appointments were on academic merit. In spite of formal emancipation, nevertheless there had been an unwritten tradition at German universities not to have more than one Jewish professor in any particular faculty. Göttingen had broken from this tradition in 1902, when Minkowski had been brought in. With Achim Gercke's work on tracing Jewish intellectuals well advanced, 1928 saw the publication of the first volume of his compilation *Der Jüdische Einfluss auf den Hohen Schulen* (The Jewish Influence on Higher Education), highlighting Göttingen. Jews were listed under seven categories, ranging from 'Mosaic', through 'Mosaic born and later baptized', to 'Jewish influenced'. With unemployment running high, one objective was to give job priority to ethnic Aryans. In spite of making up only about one per cent of the total population, in some areas, Jews made up almost a fifth of university teaching staff.[37] In Göttingen's physics and mathematics departments, it was even higher (43 per cent and 59 per cent, respectively).[38]

Born himself had been warned of his formal dismissal by a local newspaper editor, who showed him one evening the headlines of the next day's paper. Born could have qualified for exemption but preferred not to, and left Germany. Before he departed, a presumptuous colleague even asked for Born's lecture notes so that he would be able to take over the course with minimal effort![39] Courted by Lindemann, who was eager to boost Oxford's scientific reputation, Born opted instead for Cambridge, where he had been a student some twenty years earlier.[40] In May 1935 he informed his UK patrons of a growing feeling in Germany that 'emigrants had left ... as if to fight against the Nazis'. He had received a letter from the Rector of Göttingen which was 'worth seeing because of its awfully bad German', deemed by Born to be 'characteristic of these people'.[41] However it was not easy to adjust from being the head of a major department to being a visitor on a temporary grant with only occasional research collaborators. Later, after a brief stay in Bangalore, India, Born was offered a prestigious post at Edinburgh, and went on to win the 1954 Nobel Physics Prize, for work he had done while at Göttingen.

While a student at Heidelberg, Born formed a lifelong friendship with James Franck (1882–1964), who went on to become director of one of Göttingen's physics institutes. His grandparents had been religious Jews, his parents less so, but Franck remained conscious of his Jewish roots.[42] Franck played a major role in Germany's First World War poison gas development programme. Requested by his superior officers to get himself baptized, he retorted by asking them whether they thought this would make him a better officer. Although they had to admit that this would not be the case, they nevertheless said that his refusal would affect Franck's perceived loyalty to his country. He replied that his presence in the Army and their original invitation to him to become an officer should already indicate ample loyalty.[43]

After recovering from serious wartime illness, he had gone on to share the 1925 Nobel Physics Prize with his Jewish colleague Gustav Hertz. While many physicists were taking the cue from Rutherford and turning their attention to the tiny nucleus at the heart of the atom, Franck remained an atomic scientist. He worked with Eugene Rabinowitch, later to become involved in the Manhattan Project. Franck was constantly aware of his position as a German of Jewish descent, rather than a Jew who just happened to live in Germany. After the announcement of the Civil Service Law, he immediately saw its implications. His house quickly became the focus for Göttingen scientists wanting to emigrate. Distressed by how many good people around him were being dismissed, he decided that applying for exemption from the new law because of his war work meant that he was effectively collaborating with the new regime. He drafted two letters of resignation: one said that his decision to stand down as both a university professor and as director of his institute 'was an inner necessity because of the attitude of the government towards German Jewry'.[44] The other said that Jews who had served their country were being treated as enemies, an opinion which further alienated him. More than forty of his Göttingen colleagues publicly described his reaction as 'sabotage' and urged the authorities to 'speed up the necessary cleansing measures'.[45] Those, such as Born and Courant, who tried to deter Franck from leaving were also criticized.[46] Well-wishers who came to see Franck off at the railway station noted that his train was reluctant to start. As well as friends and colleagues, Franck also lost his pension.

But his scientific ability and Nobel status had attracted a lot of support for Göttingen from the Rockefeller Foundation, which was also looking to set up a new institute in Berlin, with Franck at its head. Now, this was no longer possible. After a year at Bohr's institute in Copenhagen, Franck was able to begin a new career in the United States, where Nobel Prizes attracted more prestige than they did in Germany. In the USA, Franck turned from atomic physics to biophysics, investigating photosynthesis, but was frustrated to discover that biological experiments needed more patience than those in atomic physics. After his first wife died, in 1946 Franck married his long-time scientific assistant, Hertha Sponer (1895–1968), who had been one of the first German women to qualify as a university physicist and had become a professor at Göttingen in 1932. Although neither Jewish nor politically active, she had been dismissed simply because she was a female interloper in a man's world, and had left Germany.

Hilbert's successor at Göttingen had been Hermann Weyl, formerly Einstein's collaborator, who valiantly tried to hold a tottering structure together as his colleagues left. But with a Jewish wife, Weyl's own position became untenable and he soon departed to join Einstein in Princeton. Visiting Göttingen, Education Minister Bernhard Rust famously asked David Hilbert if his mathematics department had suffered because of the departure of the Jews. 'Suffered?' replied Hilbert, 'It no longer exists'.[47] A hundred years of effort had evaporated. Göttingen was an extreme case, but everywhere students had lost their teachers, and scientists tacitly supporting the Nazis had lost the goodwill of their colleagues overseas. The free circulation of ideas so vital to the advancement of science had been checked.

Director of Physics in Berlin's *Technische Hochschule* was Franck's Nobel Prize partner Gustav Hertz (1887–1975), a nephew of Heinrich Hertz (1857–94), whose discovery of electromagnetic waves would surely have earned a Nobel Prize had he not died at the age of only 36. Franck and Gustav Hertz shared the Nobel Physics Prize in 1925 for their study of how atoms absorb electrons. Like Franck, he also had Jewish blood, but was initially exempt from the Civil Service Law because of his war service. His Nobel Prize, together with the distinguished family name, helped convince Johannes Stark that Hertz was not Jewish 'in his statements, appearance, and activities'.[48] However this leniency later lapsed, and Hertz left academia, taking up a post in the Siemens industrial organization, a rare example of a Jew who found a post outside the traditional university base. When the Soviet Army entered Germany in 1945, Hertz was one of the large team of German scientists who went to work on nuclear development in the Soviet Union, where he remained until 1955.

In theory, Fritz Haber should have been the first in line for exemption from the 1933 laws: he had saved his country in the First World War by synthesizing explosives from thin air, for which he had also earned a Nobel Prize; and he had pioneered the use of poison gas on the battlefield. Despite formally becoming Christian and looking like an archetypal German, complete with shaved head and what looked like a duelling scar on his face, Haber was still a Jew. After his first wife committed suicide, Haber married a Jewess twenty years younger than him, who converted to Christianity in order to be married to him in a church. But such window dressing would not impress the Nazis later. In 1925, Haber became a director of the mighty IG Farben chemical conglomerate.

A few weeks after the initial announcement of the Civil Service Laws, the Ministry of the Interior instructed the Kaiser Wilhelm Society that Jewish staff members were to be dismissed. For Haber, this meant three out of four department heads and five out of thirteen staff members.[49] However directors, like Haber, were exempt. With his staff decimated, Haber resigned from the Institute he had run since its establishment in 1911. In these moves, Peter Pringsheim, an expert in fluorescence and phosphorescence, also lost his post at Berlin, another case where Jewish ancestry overruled baptism. Discovering this, a furious Walther Nernst called on Haber to help reinstate Pringsheim, only to find that Haber had decided to resign himself. Pringsheim moved to Belgium. Haber received tentative job offers in England and even in Palestine,[50] but with his poison-gas war record, he had already been cold-shouldered by possible future colleagues. The dilemma resolved itself when he died on his way to Basel, Switzerland, in 1934. One year later, Max Planck organized a memorial meeting for Haber, but Nazis were instructed not to attend.

Nevertheless, the meeting was full, and the eulogy was given by Otto Hahn. Newspaper reporters attending the event were told by SS troopers not to write anything. Later, Planck, 76 years old and inured to pain and disappointment, said, 'Haber had to leave his home country for racial reasons, but the German people will have to thank him for centuries for what he has done'.[51] The Nazis, displeased, sternly told Planck not to seek re-election to the Presidency of the Kaiser Wilhelm Institute. In a letter to Haber's family, Einstein wrote that Haber's life was 'the tragedy of the German Jew, the tragedy of unrequited love'.[52]

In the 1930s, the Kaiser Wilhelm Institute for Physics in Berlin had benefited from Rockefeller funding. The continuing presence of Planck as a stabilizing factor overcame the Foundation's nervousness about the changing face of German politics. The Institute had initially been headed by Einstein, with Max von Laue as his deputy. After 1936, its Director was Peter Debye (1884–1966), a Dutchman who, apart from periods at Zurich and Utrecht, had been studying and working in Germany since 1901. In 1936 he earned the Nobel Chemistry Prize. Normally accepting such important positions in Germany would mean losing his Dutch nationality, but he had been granted formal exemption by the Queen of the Netherlands.[53] Caught up in the effort to purge German science of any remaining Jews, Debye was pressured to adopt a more pro-German stance. It was an invidious position, which made him enemies on both sides. Einstein was one of the most vociferous. Debye was also under pressure to take German nationality.[54] Reluctant to oblige, he moved to the USA in 1940, but with family still inside Germany, had to be careful what he said, even more so after Germany overran the Netherlands.

Emboldened by his acquisition of Austria in 1938, Hitler had then targeted the *Sudetenland*, German-speaking territories which had been ceded to the new nation of Czechloslovakia after 1918. In October of that year, German troops marched in to enlarge the Third Reich and extend the reach of Nazi legislation. But more trouble for Jewish intellectuals was still to come. In the First World War, Italy had fought against the Central Powers of Germany and Austria-Hungary. Although nominally victorious, it saw post-1918 upheaval as radical political ideas struggled for recognition. Of these, the Fascists under Benito Mussolini resisted modernist change, advocating instead a proud nationalism. Bands of war veterans joined the new cause, aggressively seeking out communists, socialists, and anarchists. In 1922, Mussolini's blackshirt army marched on Rome, where King Victor Emmanuel handed over governmental power. Mussolini—*Il Duce*—quickly became the world symbol of a new Fascist cause.

Flexing his muscles, Mussolini embarked on an ambitious policy of colonialism. His attack on the virtually defenceless Ethiopia was widely condemned, especially when it used modern weapons such as aeroplanes and poison gas. Italy became the victim of international sanctions which estranged it from its First World War allies. Instead, it aligned itself with its former enemy, Germany, with the official declaration of the Rome–Berlin Axis in 1936. In Mussolini's eyes, Hitler was the junior partner, whose initial antics in Bavaria had been viewed as slavish emulation of what was happening in Italy. Hitler's objectives had emerged more clearly in 1933 when Nazis assassinated the Austrian fascist leader Engelbert Dollfuss, a friend and ally of Mussolini. When Hitler annexed Austria in 1938, despite proclaimed Italian 'protection', Mussolini was humiliated.

There were not many Jews in Italy, about one in a thousand, compared with about one in a hundred in Germany, but on the other hand they could trace their history much further back: the first had arrived in the second century BC. Some were thought to be the descendants of slaves brought back from Jerusalem by the Romans. Full civil rights had been granted to them in the late nineteenth century, after Rome had been incorporated into modern Italy. Under this emancipation, and despite their rarity, by the beginning of the twentieth century Italian Jews made up 3.85 per cent of the faculty of Italian universities. By 1938, this figure had more than doubled, reaching 8.76 per cent.[55] In the early twentieth century, many central European Jews, German-speaking and aspiring to become doctors, went to Milan for their studies, where some of their examinations could even be conducted in German.[56] In 1931, all university teachers were required to take an oath of loyalty to the new regime, paralleling what was to happen in Germany (and a forerunner of what would happen again in the University of California in 1949). A few resisted, and called on their perceived saviour, Albert Einstein, to intervene with the Italian authorities. He duly wrote to Minister of Justice Alfredo Rocco and received an ambiguous reply. Although only a few had refused to take the oath, Mussolini tried to smother the resultant publicity.[57]

Although he originally had no quarrel with Italian Jews (he even had a Jewish mistress), in 1938, Mussolini suddenly introduced a wave of racial laws—*Manifesto della Razza*—to emulate those of Germany. In quick succession came Measures for the Defence of Race in the Fascist School (*Provvedimenti per la difesa della razza nella scuola fascista*) on September 5; Measures Toward Foreign Jews (*Provvedimenti nei confronti degli ebrei stranieri*) on September 7; and Founding of Elementary Schools for Young People of the Jewish Race (*Istituzione di scuole elementari per fanciulli di razza ebraica*) on September 23. While the genes of ancient immigrants were thought to have been assimilated into the Italian race, those of the Jews (*ebrei*) had not. Now, they were no longer considered to be a part of the nation. In quick succession, with a rapidity redolent of the 1933 edicts in Germany, modifications and amendments to these laws swiftly prohibited access of Italian Jews to study, whether as a student or a teacher. Immediately, many Italian Jewish scientists decided to emigrate.[58]

The most famous example was that of Enrico Fermi, of whom more later. Another case was that of Bruno Rossi,[59] to whom we will also return. On 10 October, 1938, a curt letter from the Rector of the University of Padua informed the distinguished scientist that his services were no longer needed: 'In compliance with art. 3 of the king's decree of law n. 1390 dated September, 1938, which deals with measures for the defence of race in the fascist school, it is my duty to advise you that, as of October 16, you are suspended from service'. Rossi realized that he was 'no longer a citizen of my country, and that, in Italy, my activity as a teacher and as a scientist had come to an end'.[60] In all, some 6000—12 per cent of the Italian Jewish population—had to go.[61]

NOTES

1. Schottländer, R., in *Antisemitische Hochschulpolitik*, quoted in Friedländer, p.57.
2. Letter from Selig Hecht to a US colleague, June 1933. Wiener Library, London, file 1527.
3. Schwersenz, J., *Die Versteckte Gruppe*, Wichern, Berlin, 1988, p.24.
4. Australian Academy of Science Biographical Memoir http://www.asap.unimelb.edu.au/bsparcs/aasmemoirs/lemberg.htm
5. Mussgnug, D., *Der vertriebenen Heidelberger Dozenten*, Winter, Heidelberg, 1988.

6. Cahen, F.M., *Men Against Hitler*, Jarrolds, London, 1939.
7. Hilfsverein der deutschen Juden, Wiener Library, file 602/8.
8. Cornwell, p.135.
9. Heim, p.402.
10. Friedländer, p.145.
11. US National Academy of Sciences, biographical note by George Wald, 1991.
12. Clark, R., *The Life of Ernst Chain, Penicillin and Beyond*, Palgrave Macmillan, New York, 1986.
13. SPSL archives, box 330.
14. James, p.173.
15. Kisch-Arndt, R., *A Portrait of Felix Auerbach: The Burlington Magazine.* Vol. 106, No. 732 (March 1964), p.131.
16. Friedländer, p.241.
17. Friedländer, p.264.
18. SPSL, box 330.
19. SPSL, box 342.
20. SPSL, box 329.
21. See Chapter 7.
22. Bonolis, L. and Pancheri, G., Bruno Touschek, particle physicist and father of the electron-positron collider, *European Physical Journal*, H, 10.1140/epjh/e2011-10044-1.
23. Jones, E., *The Life and Work of Sigmund Freud*, Basic Books, New York, 1981, Vol 3, p.496.
24. Waugh, A., *The House of Wittgenstein*, Bloomsbury, London, 2008, p.266.
25. Jungk, p.40.
26. Greenspan, p.174.
27. Beyerchen, p.19.
28. Waldinger, F., *Peer Effects in Science, Evidence from the Dismissals of Scientists in Nazi Germany*, Warwick preprint, 2010, Table 2.
29. Beyerchen, p.27.
30. Greenspan, p.6.
31. Born, p.6.
32. Greenspan, p.62.
33. Greenspan, p.181.
34. Greenspan, p.47.
35. Greenspan, p.154.
36. Born, p.251.
37. Dahms, H-J., Einleitung, *Die Universität Göttingen under dem Nationalsozialismus,* Becker, H. (ed), Munich, Saur, 1987, cited in Greenspan, p.165.
38. Waldinger, see note 28.
39. Szabo, A., *Vertreibung*, … , Wallstein, Göttingen, 2000, p.46.

40. James, p.202.
41. SPSL, Box 325.
42. AIP oral history, J. Franck 4609.
43. Quoted in Beyerchen, see Ash, M.G., Söller, A, (eds) *Forced Migration and Scientific Change*, Cambridge University Press, Cambridge, 2002, p.72.
44. Beyerchen, p.17.
45. Friedländer, p.50.
46. Greenspan, p.176.
47. Cornwell, p.198.
48. Swinne, p.93.
49. Friedländer, p.51.
50. Cornwell, p.139.
51. Wiener Library Eyewitness Accounts, PIIf No1129, Bernhard Landau.
52. Thiessen, V., *Einstein's Gift*, Playwrights Canada Press, Toronto, 2009, p.108.
53. AIP oral history, 4568, P. Debye.
54. AIP oral history, 4568-3.
55. Orlando, L., *Physics in the 1930s: Jewish physicists' contribution to the realization of the new tasks of physics in Italy*: Historical Studies in the Physical and Biological Sciences, Caliber, U. California Press, Vol 29, p144, and references therein.
56. AIP oral history, N. Kurti, 4725.
57. Orlando (see note 55), p.148.
58. Pekelis, C., *My Version of the Facts*, Marlboro Press, Marlboro, Vermont, 2005.
59. For a detailed account of Rossi's career, see Bonolis, L., Bruno Rossi and the Laws of Fascist Italy, *Physics in Perspective*, **13**, 2011, p.58.
60. Bruno Rossi, *Moments in the Life of a Scientist*, Cambridge University Press, Cambridge, 1990, p.39.
61. Orlando (see note 55), p.149.

7

Abide with me

The wave of Nazi enactments which began in 1933 went on to achieve what it had set out to do, changing the profile of German academe. But the simultaneous campaign of Nazi terror resulted in something else. The initial street beatings and humiliations were meant to be highly visible within Germany, but they could not be hidden from the world outside. Foreign press coverage of the violence sparked sympathy for the Jews. Although such open brutishness was soon replaced by the more discreet but also more dreadful methods of the Gestapo and the SS, compassion for the plight of Jews in Nazi Germany soon took root. This was amplified when those whose work was known, respected, or even admired, began knocking at the immigration door.

After 1933, 1145 of some 7760 senior German university teachers left or lost their jobs, mostly for racial and/or political reasons.[1] About 550 of them left the country in 1933, including about a quarter of the total physics complement.[2] Between 1933 and 1939, from *Judenboykott* to *Kristallnacht*, the teaching of exact sciences in German universities crumbled. For physics, with 325 'habilitated' teachers, some 50 emigrated.[3] By 1939, there were some 3000 fewer university teachers, researchers, and specialists working in museums and libraries, than in 1933, while student numbers overall had halved.[4] In physics and mathematics the student population decreased by 65 per cent.[5] Of the emigrant teachers, some 70 per cent managed to find at least temporary posts abroad.[6] How was such an effective job-placement mechanism set up so quickly?

Arriving in Cambridge in 1933, Max Born received a letter from Einstein, then also in Britain. 'I am glad that you [Born and James Franck] have resigned your positions. Thank God there is no risk for either of you. But my heart aches at the thought of the young ones'.[7] The initial loophole to the Civil Service Law granted exemption for war service. Anyone who served in that war was born before about 1900. In 1933, junior academics had no war record (only the minority who lost their fathers could claim exemption). Young academics are traditionally very mobile. Many leave home to go to university, then go somewhere else to complete their studies. Once qualified, they have to shift from one temporary position to another before finding a permanent job. In 1933, young university scientists were either in such short-term positions or were taking their first steps up the promotional ladder. Most of them were not yet property owners, and had yet to put down roots. Their calling was their intellect: they had no stock-in-trade or business premises to dispose of. They were prepared to go wherever there was a good job. They had recommendations from teachers, known all over the world. When exemptions were annulled, these teachers too were pushed into the emigration queue.

Only a few had seen what was coming. Einstein was one. Another was Leo Szilard. Early in 1933, his sensitive political predictor had begun to buzz alarmingly again. Just as it had told him earlier to move out of Hungary and then Austria, now it signalled that it was time to get out of Germany. He encouraged Michael Polanyi to accept a job offer that had come in from Manchester. Soon afterwards, Szilard took a train to Austria, travelling first-class, so as to appear less like a refugee, but with his baggage stuffed with rolls of banknotes.[8] He was not stopped. His predictor was extremely well tuned: the very next day, the same train was packed with Civil Service Law refugees who were interrogated at the border. On learning that the Nazis were specifically expelling Jewish academics, Szilard went to see Gottfried Kuhnwald, an Austrian Jew prominent in the Christian Social Party. Kuhnwald prophesied, 'The French would pray for the victims, the British would organize their rescue, and the Americans would pay for it'.[9] Staying at the same Vienna hotel as Szilard was the influential British economist William Beveridge, then director of the London School of Economics, and who would later provide the guidelines for the introduction of the welfare state in Britain. On return to the UK, Beveridge, seeing the value of the dismissed German scientists, went to see the nuclear physics pioneer Ernest (now Lord) Rutherford, who in the UK was probably the only scientist whose fame and familiarity as a public figure could rival that of Einstein. The quick-tempered Rutherford was furious to learn what was happening to scientists that he knew and respected. Beveridge persuaded him to chair a new Academic Assistance Council (AAC) to help the refugees, a task he dutifully carried out for the remainder of his life.[10] It was not always easy: there was high unemployment in

Britain, and there were those—particularly in medicine—who resented the prospect of increased competition in an already competitive sector. Nevertheless, those affected in Germany were surprised how quickly support came from Britain.[11] When Rutherford took on this new responsibility, Max Born at Cambridge invited Fritz Haber, who was visiting, and suggested that Rutherford should come over and meet him. In spite of his sympathy for other displaced German scientists, Rutherford refused.

After recruiting Rutherford to help raise funds for the new AAC, it was important to parade Albert Einstein, who had been staying in Belgium, and was passing through Britain before moving to the USA. Addressing a crowd of 10,000 at London's Royal Albert Hall on 3 October 1933, speaking in heavily accented English, he appealed to the 'tradition of tolerance and justice which your country had proudly upheld for centuries. It is precisely in times of economic distress, such as we experience today, that we may recognize the effectiveness of the vital moral force of a people'.[12] After those words, Einstein left Europe for good.

Meanwhile the British Cabinet had discussed the plight of the new exiles. Perhaps recalling the Anglo-German rancour which had preceded the First World War, it resolved to 'try to secure for this country prominent Jews who were being expelled from Germany and who had achieved distinction whether in pure science [or] applied science',[13] affirming that any such warm hospitality would 'create a favourable impression'. Warm words, but no action. Meanwhile the nascent AAC had begun work, taking care to use non-Jewish spokesmen. The Royal Society offered space in its London premises, which was soon occupied by key figures. One was Leo Szilard, recently arrived from Vienna, and busy writing letters to influential people on notepaper headed 'Strand Palace Hotel, London'. The others were the AAC's General Secretary, Walter Adams, formerly at University College, London, and the angelic figure of Esther (Tess) Simpson, who had been working as an administrative assistant at the YMCA in Geneva. In her new job, some 2600 exiled scholars were to pass through her capable hands. There are many testimonies to her kindness and diligence.[14] (In 1936, the AAC was renamed the Society for the Protection of Science and Learning. To avoid confusion, the initial name will be used throughout here.)

An early beneficiary was Robert Karl Eisenschitz, born in Vienna in 1898, who had acquired German nationality through marriage and was

Der Hausknecht der Deutschen Gesandtschaft in Brüssel wurde beauftragt, einen dort herumlungernden Asiaten von der Wahn-vorstellung, er sei ein Preuße, zu heilen.

Plate 1. Albert Einstein's departure from Germany.
This cartoon from the *Deutsche Tageszeitung* of 1 April 1933 says, 'The manservant at the German Embassy in Brussels was charged with curing a loitering Asian of the crazy idea that he was a Prussian'. Einstein had left Berlin for the United States in December 1932, just before the Nazis came to power. However several months later, he returned briefly to Europe, renting a house in Belgium. From there, he gave up the German citizenship he had acquired in 1914 and soon left Europe for good. The publication date of the cartoon was not accidental. It marked the official Nazi boycott of Jewish shops and businesses, the beginning of a campaign of systematic persecution. (Bildarchiv Preussischer Kulturbesitz, courtesy Réunion des musées nationaux, agence photographique, Paris)

Plate 2. *Reichstag* fire, 1933.
In February 1933, the *Reichstag*, the German Parliament building in Berlin, went up in flames. Nobody knows for sure who was responsible, but it gave the newly elected Nazis the chance to blame the disaster on others. Hitler pushed through 'emergency' legislation to enable political decisions to be taken without parliamentary approval. One of the first of these acts was the 'Restoration of the Civil Service', which purged German universities of Jewish professors and teachers. (Photo Imperial War Museum, London)

Plate 3. Nobel Prize for Physics, 1933.
In 1933, the Nobel Prize for Physics, 1933 was awarded to Erwin Schrödinger, Werner Heisenberg, and Paul Dirac for their development of quantum mechanics (Heisenberg's award being held over from 1932). None of them was Jewish. This photograph, taken at Stockholm railway station, shows (*right to left*) Schrödinger, aged 46, Heisenberg, 32, and Dirac, 31, with Dirac's mother, Schrödinger's wife Anny, and Heisenberg's mother. Schrödinger's flamboyant attire contrasts with the others' sedateness. Such was his lifestyle: Anny Schrödinger had by that time ceased to play a conventional matrimonial role in his life. Schrödinger had recently left Germany in protest against Nazi policy in general, and their treatment of the Jews in particular. In spite of his Nobel Prize, Heisenberg's career in Germany became blocked, the Nazis stigmatizing him as a 'White Jew' because of his unconventional science. (Photo Max-Planck Institute, courtesy AIP Emilio Segrè Visual Archives)

Plate 4. *Kristallnacht* 1938.
Berlin's Fasanenstrasse synagogue, built in 1912 and a symbol of the emancipation of German Jewry, lay in ruins after *Kristallnacht*, the night of broken glass, in November 1938. This pogrom marked the escalation of the Nazi persecution of the Jews: what had been discrimination now became outright attack. After seeing what the Nazis were prepared to do, refugee scientists became convinced that they had to get to the Atomic Bomb first. (Photo Imperial War Museum, London)

Plate 12. Eugene Wigner.
Eugene Wigner, born Wigner Jeno in 1902 into a Jewish Family in Budapest, came to Princeton at the same time as John von Neumann (*see Plate 8*). The world's first nuclear engineer, Wigner guided the design, construction, and operation of the huge nuclear reactors which manufactured plutonium for the first fission bombs. In 1963 he shared the Nobel Prize for Physics for work he had done before he left Europe. (Photo Argonne National Laboratory, courtesy AIP Emilio Segrè Visual Archives)

Plate 13. Jack Steinberger.
Jack Steinberger, born into a religious Jewish family in Bad Kissingen, Germany, in 1921, left the country in 1934 to continue his schooling in the United States. After becoming a leading figure in post-war subnuclear physics, he moved back to Europe in the 1960s. Science's swing westward across the Atlantic, symbolized by Einstein's departure for the USA in 1932, was beginning to bounce back. In 1988 Steinberger shared the Nobel Prize for Physics. (Photo Columbia University Nevis Cyclotron Laboratory, courtesy American Institute of Physics Emilio Segrè Visual Archives)

working at the Kaiser Wilhelm Institute for Chemistry in Dahlem. In October 1933, he was informed by the police that he had lost both his job and his German nationality. The official letter arrived soon afterwards. Help was at hand. W.H. Bragg at London's Royal Institution (who had shared the 1915 Nobel Physics Prize with his son) enquired about hiring some German immigrant talent. Leo Szilard suggested Eisenschitz, but that the job should be limited to one year, to leave the door open for others. In fact, Eisenschitz had been Szilard's second choice. The first had been Walter Gordon, fleeing from Hamburg, but he had already found a niche at Manchester. Gordon later moved to Stockholm, where, with the Swedish physicist Oskar Klein, he developed a quantum equation almost as famous as that of Schrödinger.

The AAC acted as a go-between. Most of the initial migrants were aged between 30 and 35: many already had a distinguished academic record. The AAC gave an initial grant of £250 for a married couple, or £180 for a single person, while they looked for a position, and urged universities to respond. In the first year, London University took 67, Cambridge 31, and Oxford 17, with more scattered around the country. In Oxford, Frederick Lindemann had good contacts in Germany, and had already recruited Erwin Schrödinger. With other, younger, immigrants, Lindemann turned Oxford into a centre of cryogenic expertise.

A key figure was the vice-president of the AAC and Research Professor of the Royal Society, Archibald Vivian (A.V.) Hill, who had won the 1922 Nobel prize for Physiology and Medicine for his work on muscle contraction. In a major speech, Hill underlined the need to keep politics out of science. Extracts from the speech were published in the widely read journal *Nature*, admonishing the Nazi government for its shabby treatment of such valuable science and scientists. Nazi science spokesman Johannes Stark replied that his nation had been forced to take action against 'disloyal citizens'. By the time Stark's bigoted reply appeared in print, money was rolling in, and Hill replied sarcastically, thanking Stark for his efforts in promoting the AAC cause.

The AAC became an important clearing house, matching applicants to opportunities, and providing travel money so that scientists could move on, notably to the USA. It had to tread a delicate path to attract maximum support without ruffling feathers. Not everybody had unreserved sympathy for Jews. Emigrants were helped, but the reason behind their emigration was left unsaid. However, this changed after the Nuremberg Laws. In 1935, Beveridge said, 'Events in Germany are a challenge to humanity', and were 'a relentless persecution'. 'Shadows of brutality ... are returning. ... The shadow lies deepest in Germany, because before in Germany there was the most light'.[15] 1936 saw the 550th anniversary of the University of Heidelberg, Germany's oldest, and one of the first to follow the Nazi line. Unlike Göttingen, Heidelberg did not have a high proportion of Jewish scientists in its faculty. However after vociferous campaigning by UK scientists, no British university sent a delegate to the Heidelberg anniversary ceremony.[16]

By 1939, some 900 scientists had passed through the AAC's capable hands. Half a million pounds had been collected.[17] As well as its own machinery, the AAC could draw on influential sympathizers. Patrick Blackett, who had worked with Rutherford in Cambridge before moving in the 1930s to Birkbeck College, London, and later to Manchester, was a tireless supporter of victims of the dark forces of oppression. Other mutual aid was organized by exiled Germans themselves. The *Notgemeinschaft der deutschen Wissenschaft, Kunst und Literatur im Ausland* (Emergency Organization for German Science, Art and Literature Abroad) was set up in Paris in 1934, and the *Notgemeinschaft der deutschen Wissenschaftler im Ausland* (Emergency Organization of German Scientists Abroad) in Zurich in 1933. In spite of the similarity of their names, these organizations had no direct connection with the mainstream *Notgemeinschaft* inside Germany.

As a geographical near neighbour, France was another appealing destination for emigrant intellectuals. The *Comité de placement pour les réfugiés intellectuels* took care of many, but most scientists preferred to aim for English-speaking countries. Across the Atlantic, the situation was different. Germans already had a tradition of emigration to the USA. In the early nineteenth century, turmoil in the disparate German states had made some seven million Germans cross the Atlantic. Many settled in Pennsylvania and Wisconsin, where it was not unusual to find whole areas where German was spoken. In those days, the United States had been a developing country. However, pushed by waves of enthusiastic immigrants, ample natural resources, and by a wave of technological innovation, by 1894 the USA had displaced Great Britain as the world's leading manufacturing nation. By 1914 it was producing more than most of Western Europe.[18] Around the turn of the century, a wave of invention and innovation in America went on to revolutionize lifestyle everywhere: the telephone, the photograph, the electric light, the typewriter, ... : new techniques revolutionized commerce and warfare: the telegraph, the machine gun, ... Skyscrapers equipped with automatic lifts changed the urban landscape. The leading genius of this wave of invention, Thomas Edison, also introduced the concept of the industrial laboratory, where, instead of knowledge for its own sake, applied research aimed primarily for immediate technological spinoff.

There was, however, one area where the USA had yet to make its mark: the basic science that underpins technological development. The USA had its proud universities, but they were not yet in the same scientific league as Cambridge in England or Göttingen and Berlin in Germany. The Americans knew this would have to be rectified. Their technologically based society needed scientists. In the early twentieth century, aspiring US scientists were more or less obliged to study and work in Europe to gain research experience. At one stage, there were more than twenty US postgraduates working or studying at Göttingen. One was the precocious J. Robert Oppenheimer, who earned his doctorate under Max Born in 1927. Göttingen attracted famous names, and Oppenheimer met many of them there.

A key transit centre for the exiles was Copenhagen, the home of Niels Bohr, a big, burly scientist born in the Danish capital in 1885. Bohr's father, a distinguished professor of physiology, was a strict Lutheran, a tradition Bohr duly followed as a child. His mother, Ellen Adler, came from a wealthy Jewish family: her father, David Baruch Adler, had moved to Copenhagen the year after his marriage in 1849 to found Copenhagen's Commercial Bank.[19] But Bohr's strictly Lutheran childhood must have overshadowed much of his mother's religious influence. Bohr went on to become the poet of quantum science. Before him, quantum physics had been an obscure technicality. Bohr incorporated quantum ideas into the atoms of everyday matter, showing how quantum effects dictated the way atoms shone light. As Bohr himself said: 'The poet is not nearly so concerned with describing facts as with creating images and establishing connections'.[20]

For his poetic science, Bohr was awarded the Nobel Prize for Physics in 1922. However the 1921 Nobel Physics award had been held over, and the authorities announced the awards for 1921 and 1922 at the same time. The recipient of the 1921 award was Albert Einstein. This further cemented an already strong bond between the two men, who had met for the first time in 1920, when Bohr came to talk in a still sombre and destitute Berlin, thoughtfully bringing a food parcel from Denmark.[21] Later that year, Einstein visited Copenhagen, and was charmed by Bohr's hospitality. Bohr was pleased that Einstein's Nobel award technically predated his, acknowledging the sequence of their respective contributions. Bohr's Copenhagen school became a beacon, attracting scientists from all over the world. It was there in 1927 that Heisenberg distilled the Uncertainty Principle from his new matrix mechanics.

Almost every quantum scientist called in at Copenhagen, but after 1933 it took on a new prominence, not only academically, but also as an increasingly important transit centre and haven for exiles. Vacancies and positions were posted there, and it was to Copenhagen that many of the émigrés first travelled. Most European quantum physicists worked in Copenhagen at some time. Probably all had visited there and would have met Bohr. In 1932, he moved into the impressive 'House of Honour' built by the Carlsberg brewery concern and traditionally reserved for Denmark's most prestigious citizen. The huge mansion became a material, as well as a scientific, haven, as well as a symbol of Danish national prestige. In 1934, Bohr saw his eldest son fall overboard and drown in a storm while the two of them were sailing. His grief could have sublimated into concern for scientific refugees. After the outbreak of the Second World War, Bohr remained in Copenhagen as long as he could. Initially, Denmark fared better than other countries overrun by Nazi troops. Germany technically invaded Denmark in April 1940, but with the country's rich agriculture supplying many Germans with valuable meat, cheese, and butter, the country retained much of its independence. However in 1943 the troops re-entered. Niels Bohr was now a marked man, not only for his scientific knowledge: he also had a Jewish mother.

After Albert Einstein had been recruited as the figurehead for the new Princeton Institute of Advanced Study, it looked to Europe for more talent. Its net caught two extraordinary Hungarians: John von Neumann, mathematical genius and father of computing logic and game theory, and Eugene Wigner, who went on to earn the 1963 Nobel Prize for his contributions to nuclear physics. After Einstein had earlier remarked on the generosity of his Princeton salary, von Neumann and Wigner were snapped up in a two-for-the-price-of-one move. Other US universities followed this example, seeing an opportunity to hire top talent for rock-bottom salaries.[22]

The US Rockefeller Foundation was already generous in its support for European research: it was a major contributor to the Kaiser Wilhelm Foundation and had given grants to promising young researchers working in Germany. This effort was soon channelled to help the new scientific exiles. In 1919, Stephen Duggan had founded the Institute of International Education (IIE) as a non-governmental body to stimulate educational ties between the USA and other nations. Funds came from the US government and from private sources. But as the trickle of emigrants became a stream in the 1930s, Duggan helped establish an Emergency Committee in Aid of Displaced German Scholars. It worked with museums, libraries, and universities to find employment for emigrant academics: in this way many German scholars were taken care of. It concentrated on distinguished academics whose reputations had preceded them, channelling several hundred of them towards institutions which had expressed interest. Few grants were given to institutions to support scholars younger than 35.

In 1932, the IIE's assistant director was a 24-year-old college graduate called Ed Murrow, who was soon persuaded to also become the assistant secretary of the Emergency Committee. But Murrow soon left, joining CBS as a radio reporter, where he was to leave his mark in Second World War journalism. Before the USA entered that war, several hundred displaced European scholars got jobs through the Emergency Committee. Between 1933 and 1945, the Rockefeller Foundation alone contributed $1.5 million to this cause. At the suggestion of Einstein, an American branch of the International Relief Association was formed, and subsequently merged with other initiatives to form the International Rescue Committee. With other nations now involved, the Emergency Committee in Aid of Displaced German Scholars renamed itself the Emergency Committee in Aid of Displaced Foreign Scholars. In his 1941 Annual Report, Duggan said that the work of his organization had increased 'a hundred-fold' in the past year, and compared the hospitality offered to refugee academics as 'comparable to that shown the Greek scholars by Western Europe when Constantinople fell to the invading Turks in 1453'.[23]

In the eastwards direction, one destination for German exiles in the 1930s was Turkey, where in his drastic shake-up of national affairs, Kemal Atatürk had switched the emphasis at Istanbul University from Islam towards secular studies.[24] After 1933, several hundred German professionals, including technicians and doctors, took advantage, many through the *Notgemeinschaft*. One was Arthur von Hippel, who in 1930 had become Franck's son-in-law, thereby setting himself up as a target for the 1933 legislation, despite having served as a volunteer in the First World War. Later he moved to Copenhagen, and finally settled in the USA, carrying out wartime radar development work and eventually attaining the age of 105. Others who worked briefly in Istanbul were the Jewish mathematicians Richard von Mises (who trained as a pilot in the First World War and subsequently turned his scientific attention to aerodynamics), and William Prager, who in 1932 in Karlsruhe became one of the youngest professors in Germany. Prager boldly challenged the 1933 decision to sack him and won his case in court, with the right to back pay, but did not return to capitalize on the decision. In 1940 he moved to the USA. As a German, he and his family had to travel by train from Turkey to Baghdad, by plane to India, and from there by boat to New York, a forty-day journey.

Another who went to Istanbul was climatologist Erwin Biel,[25] born in 1899 in Vienna, but who studied in Germany, becoming a *Privatdozent* in Breslau in 1932. With a non-Aryan wife, he lost this position in 1933, despite his own certificate of baptism. After seeking to join the migration to Istanbul, he worked for a brief period in Moscow, where his job became conditional on him giving up his Austrian passport and becoming a Soviet citizen, making a vow never to leave the country. Instead, he returned to academia in Vienna, only to lose his job again in 1938. Contacting the AAC in London, he said he would consider any kind of position. A supporting letter described him as 'nearly a cripple, and looks very Jewish'. On his own initiative, Biel moved to the USA, where he got a job as a meteorologist for the New Jersey Agricultural Service, which led to a faculty position at Rutgers University. In July 1939, he proudly informed the AAC that he no longer needed their support, thanking them for their help. During the war, he was a meteorology instructor for the US Army Air Force.

An alternative destination was Palestine, where the British Mandate that replaced the former Ottoman régime had also brought with it the 1917 Balfour declaration: 'His Majesty's government view with favour the establishment in Palestine of a national home for the Jewish people'. The immigration thus encouraged suddenly accelerated in 1933 with a wave of arrivals from Germany. Not all of them could find, or even hope to find, the same employment that they had been used to, but the *kibbutz* movement helped to absorb them. Such a unilateral declaration had already antagonized Palestinian Arabs, but after 1933 this ill-feeling boiled into outright rage, riots, and bloodshed, on both sides.[26] In an attempt to find a solution, a 1936 British Commission, with an eye on what had recently happened in Turkey, recommended that after the British mandate, land should be apportioned between new Jewish and Arab states, thereby creating a new dilemma.

Whatever their destination, some German Jews obtained help even before they left the country. Frank Foley, as British Passport Control Officer and Secret Intelligence Service Head of Station in Berlin, prised Jews out of internment, and Robert Smallbones, the British Consul-General in Frankfurt, worked long days (and nights) issuing entry visas on his own authority.[27] Foley also set up Paul Rosbaud in Berlin as a British agent, codename 'Griffin'.[28] Rosbaud, an Austrian by birth, had been taken prisoner by the British during the First World War, an experience which had transformed him into an Anglophile. Taking over from Arnold Berliner as editor of *Die Naturwissenschaften*, the German monthly science journal, in 1935, Rosbaud was in a good position to monitor science progress. In 1938, his Jewish wife and their daughter left for Britain, but Rosbaud stayed behind, passing information to the British on German nuclear development work, and not going out of his way to hide his disapproval of Nazi policy, showing passive resistance.[29] Wartime copies of *Die Naturwissenschaften* smuggled to Britain were monitored by Otto Frisch and Rudolf Peierls.[30]

The exodus covered a wide span of ages. Thousands of unaccompanied adolescents and children from Germany and German-occcupied lands were absorbed into foster homes in Britain and the United States through the *Kindertransport* initiative and similar efforts. Many of them would never see their parents again. During the war, some (the 'Ritchie Boys') were recruited for US intelligence work. Others went on to win Nobel Physics Prizes. Arno Penzias, born in Munich in 1933, shared the 1978 award. Jack Steinberger was born in the Bavarian town of Bad Kissingen in 1921, where his father was the cantor of the Jewish congregation. In 1934, his parents sent him to a foster home across the Atlantic: he shared the 1988 Nobel award. Bad Kissingen had a large Jewish community, and Steinberger's father taught in its school. Although the Nazi government continued to contribute towards his salary, he soon followed his son and emigrated to the USA.

From exiles to aliens

With their scientific fame preceding them, and helped by philanthropic efforts specially launched for the purpose, quantum exiles had quickly moved to new posts in Britain and the USA. However the outbreak of war brought sudden changes, first in Britain, then in the USA. Those once categorized as refugees were reclassified as 'aliens'. Although some were deemed to be engaged on work of national importance, those lower down the academic ladder were less fortunate. Haven became humiliation. By 1940, the wave of refugee-aliens had swelled with the influx from countries invaded by Germany. Worried that this could be used as cover by German spies, the British rounded up 27,000 German-speaking men. Some were transported to the Isle of Man, then to Canada or Australia, where they were interned and treated like prisoners of war.[31]

One victim of this treatment was Walter Heitler, who had worked with Erwin Schrödinger in Zurich before qualifying as a university teacher at Göttingen under Born. In 1933, Born was able to get Heitler a junior post in Bristol, where he worked for a time with co-emigrés Hans Bethe and Heinz London. In Bristol, Heitler wrote his magnum opus *The Quantum Theory of Radiation.*[32] which would remain a standard work for the next ten years. Heitler's grant allowed him to work wherever he wanted, but an enquiry about a move to Cambridge was turned down by the considerate but pragmatic Rutherford, who pointed out that there 'some resistance to "them" all coming to Cambridge', and that expatriate German scientific prowess should be distributed around various universities.[33] Heitler's lecturing at Bristol was much appreciated. In 1936, students who were facing a major exam asked Heitler to continue his lectures on modern physics, even though the material was beyond their exam syllabus.

In 1937, Heitler travelled to Berlin for medical treatment. Crossing the border at Aachen, his passport was taken away and he was told to report to the *Gestapo* on arrival in Berlin. Heitler was surprised to discover that the *Gestapo* knew about his career in Britain, even that he had just published a major book. His passport was returned, but he was told that, in future, permission to return to Germany should first be requested from the Germany Embassy in London. Heitler had just begun the tedious process of applying for British naturalization when on 26 June 1940, a police car came to take him away to an unknown destination. Heitler's predicament had not been helped when Mrs Heitler had washed her husband's light-coloured trousers and hung them out to dry. This had been interpreted as a possible signal to German agents. After release from internment, Heitler worked with Schrödinger in Dublin.

The police car that came to take Heitler away also took Heinz London and the extrovert Herbert Fröhlich. Born in Rexingen in 1905, Fröhlich was a keen sportsman and had no hesitation about openly describing himself as Jewish. A student of Sommerfeld, he became a *Privatdozent* in Freiburg in 1931, only to lose the position two years later. He threw his Civil Service Law letter in the waste basket, but retrieved it to make an important calculation on the back, which he kept. But even before 1933, the Fröhlich family had seen that trouble was brewing. After Herbert's younger brother had been arrested by the SS for circulating communist literature, his father convinced the police (yet to come under Nazi control) that they were better suited to handle the affair. The boy was released after one night in the cells, but the father was subsequently beaten up by the SS.

Just before the Nazis came to power, Fröhlich described an encounter with a fellow student:

> STUDENT: 'The Jews have decided to destroy Germany.'
> FRÖHLICH: 'How do you know?'
> STUDENT: 'I've read the *Protocols of the Elders of Zion.*'
> FRÖHLICH: 'Do the Jews know this?'
> STUDENT: 'Of course.'
> FRÖHLICH: 'Well, I am a Jew and I do not know that.'
> STUDENT: 'You must be an exception'.[34]

Imaginatively, Fröhlich moved from Freiburg to Leningrad. He escaped from Germany by getting off his train as the SS guards got on, just before it crossed into Switzerland at Basle, and then walking across the frontier. As a German in the Soviet Union, he soon encountered Stalin's Great Purge. In Leningrad, he regularly had to leave his passport with the police. On one occasion, he retrieved it decorated with a Soviet exit visa valid for five days, four of which had already elapsed. Travelling via Vienna, his next port of call was Palestine, then administered by the British, where strings were pulled to obtain a post for him. But in this game of academic musical chairs, he preferred to go elsewhere. A possible move to Bristol was delayed because Klaus Fuchs was already there. After his internment was sorted out, Fröhlich returned to Bristol, eventually moving to a professorial chair at Liverpool in 1948 at the invitation of James Chadwick.

The treatment of the interned junior academics was in contrast with their later achievements. Examples were Max Perutz, born in Vienna in 1914, who had just completed his Cambridge doctorate work on the molecular structure of haemoglobin, and fellow Austrians Thomas Gold, Hermann Bondi, and Walter Kohn. After pleas from British scientists, these young scientists were released from internment and reinstated. Perutz went on to share the 1962 Nobel Prize for Chemistry for his work on the structure of protein molecules. Kohn shared the Chemistry Prize in 1998 for the elucidation of the electronic structure of materials. Gold and Bondi became distinguished cosmologists. Among his other roles, Bondi also became chief scientific advisor to the UK Ministry of Defence. However, the interned aliens subsequently released also included the German physics student Klaus Fuchs, who went on to play a distinguished role—as a nuclear spy and a traitor. There will be more on him later.

There was a similar sordid incarceration of German immigrants in the USA, including some extradited from South America. Some were covered by ad hoc exchange agreements negotiated through the International Red Cross and were repatriated to Germany, even while the war was still on. They had not anticipated returning to ruined cities under continual air attack, a stark contrast to life in the USA. The unsympathetic German authorities, suspecting that the returning emigrants could harbour spies, promptly interned them, where some remained for the duration of the war.[35]

'Carrying stones for ten hours each day'

Even for those scientists stranded inside Germany after the outbreak of war, there could be help. Although Jewish and an exact contemporary of Einstein, Richard Gans had a very different career. Einstein himself said that Gans 'was one of the most prominent and multiply talented physicists ... under difficult circumstances and in completely unlearned surroundings'.[36] By 1943, Einstein had been at the prestigious Institute for Advanced Study for ten years. The world's leading scientist had just agreed to be a consultant on explosives for the US Navy. That same year, Gans, who had only recently been doing consultancy work for German industry, was given a very different assignment. In March 1943, aged 63, the small, fragile man who had never had to do any hard physical work in his life was ordered by the Nazi *Amt für Judeneinsatz* (Jewish Employment Agency) to shift rubble on a Berlin bomb site. It was softening-up for a concentration camp: three months before, the Nazis had begun deporting Jews inside Germany to Auschwitz-Birkenau.

Born in Hamburg in 1880, Gans came from a comfortable family of established Jewish merchants who, unlike the Einsteins, did not need to be continually on the move. In those days, there were many Jews in Hamburg. Although nominally assimilated, they had become noticeable by their sheer numbers. In the Wilhelm Gymnasium (also attended by James Franck, two years younger than Gans), about a quarter of the students were Jewish, which led to allegations of it being a '*Judenschule*'. This seemed to sensitize Gans, who initially described himself (as Einstein had done) as '*mosaisch*' rather than Jewish, before advertising himself as '*konfessionslos*' (without any confession) to avoid further obstacles.

After initially studying engineering, he switched to physics, doing research work first at Heidelberg, before moving to Tübingen in 1903. Again, his Jewishness became visible. Pushing through his *habilitation* candidature involved his teachers in a considerable effort, even though 'he seemed to have no use for his religion'.[37] After publishing a book which became a standard text on the mathematical technique of vector analysis, and with a series of learned papers on magnetism to his name, he moved to Strasbourg, before emigrating in 1912 to the University of La Plata, Buenos Aires, Argentina, establishing there a new institute of physics. He was not the first German scientist to go South America. With the geographical extent of its empire dwarfed by those of Britain and France, Germany exported its culture to help increase its visibility.

Thus Gans was absent from Germany throughout the FirstWorld War, in which two of his brothers died. He was also away when great changes were happening in science. He returned to his home country in 1925 to become professor at Königsberg's Albertus University. The institution had a great tradition, particularly in mathematics: Hilbert and Minkowski had taught there. Helmholtz had been one of its distinguished physics teachers. In 1932, Gans' wife died, and this setback was underlined the following year by the cold-water treatment of the 1933 Civil Service Law. In spite of him not advertising his background, a Königsberg internal report, before even mentioning his scientific achievements, began 'Gans is the son of a Hamburg wholesale merchant, who was a Jew. He himself is a dissident'. Gans was described as having used 'sharp words' in his political opinions.[38] The report goes on to note that, in distant Argentina, Gans could take no part in the War, but did not hurry back afterwards, preferring to remain in South America and have his war loans—worthless inside Germany—fully reimbursed, so that he could 'grieve in solid gold' for the loss of his brothers in battle.

But unlike many of his contemporaries who fled the country, Gans was initially able to resist the Civil Service Law. The University Rector dared to write to Berlin, pointing out that Gans was 'scientifically excellent' and a 'good teacher'. Gans' former colleagues in Argentina also supported him. As the pressure increased, a further line of defence was a remarkable letter from no less than Johannes Stark, the high priest of *Deutsche Physik*, who had already acted to help Gustav Hertz. While admitting that Gans was not in the same class as Hertz, Stark pointed out that his work nevertheless had scientific value, and that he had 'kept himself distant from the Einstein set'.[39] However despite such remarkable support, and perhaps because of Stark's waning influence, it did little to help, and on 14 October 1935, Gans was officially 'retired'. Now unable to enter the establishment of which he had recently been director, he could not hold a farewell seminar, or even use the library.

Shut off from academia, he moved to Berlin in 1936. Reclassified as a 'privileged non-Aryan' he was able to get a job as physics consultant at the research laboratories of the AEG electrical engineering concern. Here, he met others who had similarly been driven from university positions, but who nevertheless managed to hold a job, as well as resisting the continuing Nazification of German physics. It was not a comfortable life, and in 1937 the stressed Gans underwent an operation for a stomach ulcer. The outbreak of war found him doing laborious calculations on radar transmission, but he later moved to fission work with Manfred von Ardenne. No longer classified as privileged and wearing instead the yellow Star of David badge denoting that he was a Jew,[40] his usefulness to the German cause, indeed his own fate, now hung by a thread.

Heinz Schmellenmeier was a scientist and an active communist, arrested in 1936 but released after five months. Unable to get a job and unwilling to serve as a soldier, he ingeniously created his own modest laboratory, carrying out contract research funded by the army ordnance office. Just before midnight on 25 March 1943, he got a phone call from Fritz Houtermans, working in the fragmentary German fission development effort, and who knew from bitter experience in the Soviet Union how unpleasant life could become for dissidents. Houtermans reported that Gans, at the age of 63, was having to 'carry stones for ten hours each day' on Berlin bomb sites.[41] He was being broken in preparation for a worse fate.

A network of scientific supporters sprung into action to convince the authorities that Gans, an expert in magnetism, would be more profitably employed in Schmellenmeier's laboratory. A letter from no less than SS *Reichsführer* Heinrich Himmler appeared to settle the matter. The enfeebled Gans was assigned a new objective.[42] More than a decade before, an obscure idea in a German research journal had led the US scientist Ernest Lawrence to invent cyclotrons—machines to whirl protons and other subatomic fragments and accelerate them to high energies. Schmellenmeier's idea was investigate whether these new machines could be made into weapons. The effectiveness of conventional German anti-aircraft guns had been reduced by Allied bombers flying high, out of range of the guns, and wreaking increasing havoc on German cities. Perhaps Schmellenmeier's new weapon could hit such targets in the stratosphere. The idea was to whirl electrons round in a ring to create intense microwave radiation. This would then be beamed at enemy aeroplanes and interfere with engine ignition or blind the crew. The device was called the 'Rheotron'. On paper, it sounded very impressive, a new superweapon to complement the V1 and V2 *Vergeltungswaffen* (retaliation weapons) then being readied. However, the idea of such a 'death ray' was now new. In 1935, Britain had set up a committee to study new methods of air defence. Soon, it appeared that the beam power needed was unattainable, but the technology was quickly refocused into what became known as radar.[43]

Gans had been rescued once from fate in a concentration camp, but was still not safe. His luck finally seemed to have run out in the summer of 1944, when he was arrested and taken to the Grosse Hamburger Strasse, Berlin's assembly point for Jews en route to concentration camps. Again, a rescue mission swung into action at the last minute. By now, with the fate of the Nazis becoming clear, some high-level SS members were eager to fabricate some protection for themselves by helping Jews instead of killing them. The Rheotron project was an elaborate bluff on several levels. First, it was a paper deterrent, but it also had a secondary role as a cover for Jewish scientists, notably Gans, who suspected that the idea would not work.[44] Later that year, Schmellenmeier's equipment was evacuated, with the instruction that 'in the case of military defeat, the Jew Gans is to be liquidated'. However, the advancing American forces arrived first.

But Gans' problems were not yet over. His two sons were arrested by the Soviet occupation authorities and deported. Gans himself was also thinking of the future of German science and its fine universities, blighted by the exposure to *Deutsche Physik* and the war. One survivor of the great days of German science was Arnold Sommerfeld, who tried to entice Gans to Munich to help put the university on its feet again. Other universities also clamoured for his help, pointing out the generous daily ration allowances for professors—100 grams of meat and 30 grams of fat.[45] However Gans had other plans. Although officially forbidden to leave Germany, he surreptitiously moved to Paris. In his pocket was a letter from Einstein, advising him where to go next.[46] Gans took the note, and soon reappeared in Argentina, where he was reinstalled as Director of the Physics Institute of the University of La Plata. Perhaps he should have stayed there in 1925. When he died in 1954, one year before Einstein, his colleagues in Argentina wrote: 'he is survived by students and colleagues in countries that he so faithfully served and which treated him so badly'.

NOTES

1. Noakes and Pridham, pp.249–50.
2. Medawar and Pyke, p.29, Noakes and Pridham, p.250.
3. Ash, M. and Söllner, A. (eds), *Forced Migration and Scientific Change*, Cambridge University Press, Cambridge, 2002.
4. McDonough, p.47.
5. Beyerchen, p.171, quoting Hartshorne, E., Numerical Changes in the German Student Body, *Nature* **142**, July 1938.
6. League of Nations report, 1934, cited in Niederland, *The Emigration of Jewish Academics and Professionals from Germany in the First Years of Nazi Rule*, Leo Baeck Institute, New York, 1988.

7. Born, M., *The Born-Einstein Letters*, Macmillan, New York, 1971, cited in Rhodes MAB, p.192.
8. Lanouette, p.116.
9. Rhodes MAB, p.192.
10. Campbell, p.457.
11. Teller, E., *Memoirs*, Perseus, Cambridge MA, 2001.
12. Pais ELH, p.195.
13. Cabinet papers, cited in Medawar and Pyke, p.54.
14. SPSL archives.
15. Zimmerman, D., *Minerva* **44**, 25–45, 2006.
16. Zimmerman (see note 15), p.43.
17. Zimmerman (see note 15), p.30.
18. Bryson, B., *Made in America*, Secker and Warburg, London, 1994.
19. Pais NBT, p.39.
20. Quoted in Cropper, W., *Great Physicists*, Oxford University Press, Oxford, 2003, p.250.
21. Pais ELH, p.40.
22. Weisskopf, Chapter 10.
23. Strauss, Emigration, *Teil* III.
24. Beyerchen, p.29.
25. SPSL, box 324.
26. Gilbert, M., *Jerusalem in the 20th Century*, Chatto and Windus, London, 1996, p.148.
27. Many more were to help, for example Oskar Schindler. The *Yad Vashem* memorial in Jerusalem lists some 22,000 'Righteous Among Nations' – non-Jews who ran risks to help Jewish victims of Nazism.
28. Kramish, A., *The Griffin, The Greatest Untold Espionage Story of World War II*, Houghton-Mifflin, New York, 1986.
29. Jungk, pp.96–7.
30. Williams, p.44.
31. The experience was eloquently described by Max Perutz in the *New Yorker*, **12** August 1985. The article was reprinted as an Appendix in Medawar and Pyke.
32. Heitler, W., *The Quantum Theory of Radiation*, Oxford University Press, Oxford, 1936.
33. SPSL, Box 330.
34. SPSL, Box 328.
35. German TV programme, 2009.
36. Swinne, p.60.
37. Swinne, p.13.
38. Swinne, p.61.
39. Swinne, p.93.

40. Waloschek, P., *Todesstrahalen als Lebensretter*, Books on Demand GmbH, 2004, Chapter 3.
41. SPSL, Box 330.
42. Waloschek, P., *Todesstrahalen als Lebensretter*, Books on Demand GmbH, 2004, Chapter 3.
43. Jones, R.V., *Most Secret War*, Hamish Hamilton, London, 1978, p.42.
44. Waloschek, P., *The Infancy of Particle Accelerators*, Vieweg, Wiesbaden, 1994, p.76.
45. Swinne, p.135.
46. Swinne, p.143.

Fission mission

When Albert Einstein was first told about the prospect of nuclear energy by Leo Szilard, the architect of relativity perceptively observed that this would be the first time that mankind could draw on a source of power which did not ultimately derive from the Sun. Szilard had explained to him that atomic nuclei could splinter apart in a self-sustaining chain reaction.[1] If this chain reaction could somehow be controlled, nuclear power could be tapped. If not, the new energy source would produce a terrestrial explosion of unprecedented power.

The signpost which had first pointed to this route was radioactivity, the natural break-up of unstable nuclei. This was another of the scientific novelties that appeared around the turn of the twentieth century. Curiously, it advanced with the help of women scientists at a time when such talents were largely neglected. In those days, women were supposed to be housewives and mothers. In Germany, their motto was *Kinder, Kirche, Küche*,—children, church, and kitchen. In 1895, New Zealand became the first country to give women the right to vote, but other nations did not hurry to follow this example. In the early twentieth century, women with intellectual ability and ambition usually became schoolteachers or governesses. Anything else required fortitude and fortune. The study of radioactivity was an example: even the name had been invented by a woman—Marie Curie. Suddenly, aspiring woman scientists had a role-model.

Marie Sklodowska-Curie, the first woman to win a Nobel Prize (for physics in 1903 for her work on radioactivity), and the only woman to have two Nobel Prizes (also that for chemistry in 1911), was a scientific heroine. Born in Poland in 1867, she initially worked as a governess, but her scientific ambition took her to Paris, where she convinced her professor, Pierre Curie, that radioactivity was the way to go. They got married into the bargain. When Pierre was killed in a Paris street accident in 1906, his wife was left alone with children and with an enormous responsibility to science. She inherited her husband's post, becoming the first woman professor at the Sorbonne, and in 1911 was awarded her second Nobel Prize for her discovery of radium. Her daughter, Irène, went on to continue the tradition, sharing the 1935 Nobel Chemistry Prize with her husband, Frédéric Joliot.

Lise Meitner also achieved scientific fame with radioactivity, but while Marie Curie had overcome adversity in her youth, Lise Meitner became a victim of it as she grew older. Early photographs show an attractive woman, fashionably dressed and self-possessed. Later ones from scientific conferences show her seated in the front row alongside grim Nobel laureates, the only woman in a room full of famous men, and demurely, often dowdily, dressed. Sitting in the front row of such photographs was already a major accomplishment: it did not need to be underlined by a fashion statement. In later life, she looked tired and sad. This was understandable.

Born in Vienna in 1878, Lise Meitner was the third of eight children in a distinguished family of Jewish intellectuals.[2] Her father was among the first Jews in Vienna to study law, and her elder sister Auguste became a composer and pianist. Although they lived among Jews, they distanced themselves from Jewish life to such an extent that Lise Meitner knew little of it. With adequate secondary schooling for bright girls limited, she was tutored at home in her favourite subjects—mathematics and physics. At Vienna University, she was among the last students to be taught by Ludwig Boltzmann; deeply motivated by his lectures, she was devastated by his suicide. With his disappearance, research at Vienna turned to a new branch of science. It would become Meitner's lifetime mission.

Unstable nuclei are like subatomic volcanoes. Their eruptions throw off fragments, large and small, until they finally quieten down. Just as terrestrial volcanoes emit molten lava as well as solid pieces of rock, so radioactive nuclei vomit a smooth stream of electrons as well as hard nuclear fragments. Such nuclear transformations tread a borderline between physics and chemistry: the processes at work inside the nuclei are physics, but a nucleus losing a part of itself transfers to a new place in the Periodic Table of elements. Such transfers are chemistry.

After she earned her doctorate in physics in 1905, one of the first women to do so in Vienna, Lise Meitner's ambition led her to Berlin, there to learn from another magisterial figure, Max Planck. The instigator of the quantum idea was kind and hospitable, and she enjoyed many visits to the Planck home. But university life was a different matter: as a woman she was neglected and sidelined. In spite of his domestic hospitality, Planck could only offer her a meagre assistantship. The rules did not allow any formal role for a woman. However, at Berlin Lise Meitner soon met the man who was to be her longtime scientific partner. Otto Hahn (1879–1968) had just returned to Germany after pioneering work on radioactivity with Ernest Rutherford in Montreal. Hahn was supremely skilled at the chemical analysis needed to track what happened in radioactive decay, but wanted a physicist to help him to see its nuclear implications.

The Berlin chemistry group was headed by Emil Fischer, who had won the 1902 Nobel Chemistry Prize for his work on sugar synthesis. Not only did Fischer not want women in his team, but he did not even want Lise Meitner to be seen in the laboratory 'in case her long hair caught fire in a Bunsen burner'[3] (apparently men's beards were more flame-resistant). As a hard-won concession, the Hahn–Meitner collaboration could continue as long as it was relegated to a carpentry workshop in the basement, with its own entrance. From there, Lise Meitner initially had to go to a café down the street to use the toilet. While the polite Hahn went out of his way to be correct, Lise Meitner's other male contemporaries showed a disdain for a female academic which was in total contrast to the gallantry and courtesy they would have extended to any woman in a more conventional setting. The start of the Hahn–Meitner collaboration was not easy. Hahn describes how Lise Meitner was reserved and shy, which he attributed to her family upbringing,[4] but was probably just as much a reaction to her icy reception in Berlin. Certainly she did not like to bring herself to people's attention: like many of her family, she was an accomplished pianist, but preferred to reserve this for family occasions. While Hahn regularly met his Berlin colleagues over lunch, or in coffee houses, for years his only contact with Lise Meitner was in their laboratory, with one or other of them going out to get bread and cheese, or sausage, but which they never ate together. For a long time the personal exchanges between them were proper, even prudish. For sixteen years they called each other by their titles and family name.[5]

In spite of this distance, their scientific work was fruitful. After a break during the First World War, when Hahn joined the German poison-gas effort, and Lise Meitner volunteered to accompany the Austrian army as an X-ray assistant, they worked together in Berlin on and off for the next thirty years. Conditions soon improved at a new Kaiser

Wilhelm Chemistry Institute in Dahlem, on the outskirts of the city, where Lise Meitner was officially put in charge of a physics group within the Chemistry Institute, with a salary. People now spoke of the Hahn–Meitner laboratory. After the war, Hahn and Meitner discovered a new radioactive element, number 91 in the Periodic Table, which they called protactinium. For more than a century, the discovery of new elements had been one of the great games of chemistry. By 1918, the Periodic Table was almost completely filled, and protactinium sealed one of the few remaining gaps.

A woman in a man's world, Lise Meitner scaled the academic promotion ladder slowly, first moving from tolerated hanger-on to paid scientific associate, then in 1922, at the age of 44, progressing to formal *habilitation*, the final step towards qualifying as a university teacher. In 1926 she became an assistant professor, one of the first women in the country to achieve such status.[6] She taught a course with Leo Szilard.[7] For a quarter of a century, she devoted herself to her work, and became a world authority on radioactivity: Einstein described her as the 'German Madame Curie' (inaccurately, as she was Austrian).[8] When in 1930 Wolfgang Pauli sent a letter apologizing for his absence when invited to speak at a special meeting, it began 'Dear Radioactive Ladies and Gentlemen'. It should have said 'Dear Lady and Gentlemen': the only female was Lise Meitner.

The 1933 Civil Service Law suddenly wiped out Lise Meitner's hard-won position at the university. Her claim for exemption because of her wartime medical work was refused, as she had been judged not to have been at the front. However she still had her position at the Kaiser Wilhelm Institute, in which, as a non-governmental organization, she was initially exempt from the restrictions. Here, she immersed herself in her science, which had become her life. There was very little outside of it. However her correct behaviour with the German chemist had gradually warmed over the years: she now called Hahn ('cock' in German) the affectionate diminutive *Hänchen*.

One of her research assistants in Berlin was Max Delbrück (1906–81). After making seminal contributions to nuclear physics (now known as 'Delbrück scattering'), in 1937 his lack of enthusiasm for Nazi politics became noticed, and he left for the United States. There, he switched to the burgeoning field of molecular biology—a career move apparently initially suggested by Niels Bohr—and went on to become a key figure in this new field of research. He earned the Nobel Prize for Medicine in

1969 for showing how bacteria become resistant to infection. Another colleague of Lise Meitner in Berlin was the botanist Elisabeth Schliemann (1881–1972), whose teaching position was later revoked after she questioned the warped Nazi view of science.

Lise Meitner either did not realize the implications of the political developments in Germany, or chose to ignore them. She turned down offers of a year at Bohr's Institute in Copenhagen, and invitations from US colleges.[9] She was now nearing 60, old by the standards of the time, and considered that it would have been difficult to take emigration in her stride. It would also interrupt her work. As an Austrian, Lise Meitner had initially sidestepped the 1935 Nuremberg race laws, but with Hitler's annexation of Austria in 1938, she technically became German, and all that that now implied. Although she had been baptized, her Jewish ancestry meant that she was now in danger of losing her position at the Kaiser Wilhelm Institute. Hahn, taken by surprise, went to the institute's authorities, and to representatives of the industry which funded it, to seek exception for his research partner. Not only was the answer an emphatic 'no', but he was told that she had to go immediately.[10] Hahn warned her to stay away from the laboratory that had been her working life. The Civil Service Laws were now being adjudicated on a case-by-case basis, so there were sometimes conflicting messages. There were signals that the Nazis had decided that enough university teachers had left, and that no more—Jewish or otherwise —would be allowed to leave.[11] Loud voices within the Kaiser Wilhelm Society had shouted 'The Jewess endangers the Institute'.[12] Now the same voices warned the police that Lise Meitner might try to slip away.

Invitations to give lectures in Zurich and Copenhagen offered Lise Meitner a chance of escape, and James Franck in Chicago took the first steps towards offering her a haven in the USA.[13] Her fears that she would both lose her remaining job and be prevented from leaving the country seemed to be confirmed by a formal letter from the Ministry of the Interior.[14] After unsuccessful high-level representations to the Kaiser Wilhelm Society, she turned to Peter Debye, the head of the nearby Physics Institute. Through contacts in Zurich, Debye alerted his Dutch colleagues, one of whom was despatched to Berlin to fetch her.[15] Help also came via Paul Rosbaud. Lise Meitner slipped away from Berlin almost unnoticed. Hahn was one of the few who knew. He gave her a diamond ring, a family heirloom, as an insurance policy: her Berlin bank account would not be accessible from outside Germany.

She chanced taking the train to Holland, pretending to be going on vacation.

Heading for Groningen in the Netherlands, she crossed the border near Nieuweschans, a small railway station at the eastern most point in that country. There are conflicting accounts about what happened. In a commendable biography which used Lise Meitner's diary, Ruth Lewin Sime says it went 'without incident'.[16] Another version alleges that she was 'so frightened, [her] heart almost stopped beating' when German guards came onto the train. Her only travel document, her invalid Austrian passport, was taken away for examination. The guards came back after some time and returned it to her, without comment.[17] On the Dutch side, there was no problem: she was expected. (The biography by Patricia Rife contains a similar report of what happened that day. Prominent international scientist Herwig Schopper is one of the few scientists still living to have known Lise Meitner, having worked with her in Stockholm in the early 1950s. As a German, he frequently spoke with her about her experiences in that country, but she never mentioned her departure from Berlin.)[18]

In June 1938, the AAC in London had been told that a passport would only be issued to her if she 'obliges not to return to Germany': in other words, she would have to emigrate.[19] Even that would be difficult. In 1938, exit visas for German Jews were granted only if all of the detailed regulations had been scrupulously followed. Bank accounts were monitored, so it was difficult for refugees to accumulate wealth by selling off their possessions. Their only option was to sell their belongings cheaply for cash. Then came emigration tax (*Reichsfluchtsteuer*), originally imposed at 25 per cent of transfers worth more than 200,000 Reichsmarks (about 48,000 dollars), but this cutoff was soon reduced to 50,000 RM (12,000 dollars). With other punitive impositions, the tax rate was gradually increased to a painful 80 per cent.[20] There was strict control at frontier posts, and travellers with baggage were only allowed across if all their papers were in order. This could be a harrowing experience.[21] However there was an escape clause: *Kleinverkehr* (small traffic) allowed travellers to circulate within 40 kilometres of the frontier, provided that they carried only ten Reichsmarks ($2.40).[22] Such travellers would not be carrying much baggage. On that day, Lise Meitner—60 years old, with an invalid passport, travelling alone, at an obscure border crossing, looked as though she fitted into that category. During the journey, she had handed Hahn's ring to the Dutchman who was accompanying her. Her purse contained the prescribed ten marks, nothing more.

When she arrived in Groningen, she sent a coded telegram to Hahn: the 'baby' had arrived.[23] After a short stay in the Netherlands, she transited through Bohr's haven in Copenhagen. Her next objective was to be able to enter to Sweden, still with her effectively invalid Austrian passport. After this hurdle had been crossed,[24] and with funds from the Nobel authorities, Lise Meitner was given a niche in the Stockholm laboratory of 1924 Nobel Physics laureate Manne Siegbahn. Although safe, she was nevertheless a sad figure, with no money and few clothes, dependent on others for support. At her age, most people were looking towards retirement, not seeking a new job. Meanwhile the Nazis had blocked her pension entitlements and conducted a detailed inventory of her belongings. Her remaining clothing, dutifully sent by Hahn, did not arrive until several months later. Not speaking Swedish and eight years older than laboratory head Siegbahn, she was distanced from her new colleagues. She had also been cut off from the science that had been her life. After the Nobel Prize festivities in Stockholm in December 1938, Laura, the wife of the new physics laureate Enrico Fermi, described Lise Meitner as looking worried, tired, and tense.[25] Although her family in Vienna was now even more distant, she cherished one family connection. In Berlin, she had regularly welcomed at Christmas her nephew, the son of her musically talented sister, Auguste. His name was Otto Frisch. This time he would come to Sweden rather than Berlin, and together they would change the course of science.

If a child is gifted, it does not know. Nor do its parents at first. They have to discover. For the family of Otto Frisch, the penny seems to have dropped when he was about five: he could handle the arithmetic of fractions at a time when most children are still learning to count.[26] Like the Meitners, the Frisch family had abandoned Judaism as part of their lives by the time Otto was born in Vienna in 1904. His father's family had a tradition of commercial printing: his grandfather had introduced to Vienna the idea of printed forms with blank spaces where information could be inserted by hand. His father, now married to Auguste Meitner, eventually sold his printing business, but as a hobby compiled encyclopaedias. He introduced his son to higher mathematics as best he could, but Otto continued his tuition with Olga Neurath, a blind mathematical virtuoso. Mathematics was his individual talent, but the entire Frisch–Meitner family had a passion for music. Otto Frisch was an enthusiastic pianist and violin player. In the days before radio and recording, musical skills were a useful social accomplishment. But, unusually for a scientist, Frisch was also a reasonable artist: his charming but brief autobiography *What Little I Remember* is enlivened with cartoon sketches of his contemporaries.

In science, Otto Frisch had practical as well as mathematical abilities. After graduating from Vienna University in 1926, he worked for a short time with an electrical company, making X-ray equipment and the like. By this time, his aunt had a position in Berlin University, and she persuaded him to move. On his arrival in the German capital in 1927, Frisch had to accustom himself to several things: the Berliners' way of speaking German, and traffic driving on the right. At that time, Austria still drove on the left.[27]

Initially, Frisch worked in the optics division of the *Physikalisch Technische Reichsanstalt*, and would occasionally go to physics seminars at Berlin University. Already impressed by the heavyweight line-up in the front row of these events, he was inspired when his aunt introduced him to Einstein. The great scientist waited patiently with his hand out while Frisch transferred a pile of books from his right hand to his left.[28] As yet, Frisch had little science to offer, and remained in the background, unlike his contemporary Leo Szilard. But his exuberant piano playing became popular, and was one of the few areas where he felt he could be extrovert among his seniors. However, he was more disciplined when playing duets with his aunt. After two years, he moved on to part-time work in the university proper. In 1930, he graduated to his first full academic post, this time at the University of Hamburg, as assistant to Otto Stern, who with Walther Gerlach had made key measurements of atomic magnetism. In Hamburg, Frisch enjoyed making precision measurements of the quantum effects of electrons. To relax after the discipline of laboratory work, he and his friends would play chamber music together, or drive across to the *Reeperbahn* red-light district.

Like his aunt, Frisch had not been paying much attention to political developments, and as a foreigner at first did not attach much importance to Hitler's sudden appearance on the political stage. Then came the Civil Service Laws. Frisch, who only had a temporary position in Hamburg, had obtained a US Rockefeller Foundation scholarship to work in Rome with Enrico Fermi, but the award was conditional on him having a German job to come back to. As this could no longer be the case, the scholarship offer became invalid. Otto Stern, also a Jew, had been cast adrift by the Civil Service Laws. However, the influential and resourceful Stern soon arranged a job for himself at the Carnegie Institute in Pittsburgh. Later he would be awarded the 1943 Nobel Physics Prize. For his assistant, he arranged a one-year position in London's Birkbeck College to work with Patrick Blackett. A disciple of Rutherford, Blackett went on to earn the 1948 Nobel Prize for Physics. During the 1930s, he worked hard to help the oppressed and underprivileged, and was delighted when Stern enquired about placing his protégé at Birkbeck.[29] In the few months before finally leaving Hamburg, Frisch heard and read many reports of Jews being beaten up and was wary of going out after dark. When he did run into a burly youth in Nazi uniform late one night, it turned out to be his landlady's son, who was very friendly. In London, Frisch was surprised by an apparent lack of organization which contrasted with German thoroughness. Instead of drawing supplies from well-equipped laboratory stores, students had to go to Woolworth's to buy bits and pieces for their experiments. After his year in London, Frisch moved to Bohr's group in Copenhagen.

Chain reactions

In Paris in 1934, the husband and wife team of Frédéric Joliot and Irène Curie synthesized 'artificial' nuclei that were also radioactive. For this discovery, they were awarded the 1935 Nobel Chemistry prize. In his Nobel Lecture, Joliot pointed out that the new development suggested that self-sustaining nuclear 'chain reactions' could be produced with 'enormous liberation' of energy.[30] In chemistry, the implications of such chain reactions had been pointed out earlier by Michael Polanyi.

The discovery of the neutron, by James Chadwick in Cambridge in 1932,[31] opened up a new nuclear front. Carrying no electric charge, neutrons could easily penetrate the intense electrical concentration of the atomic nucleus. 'When a neutron enters the nucleus, the effects are about as catastrophic as if the Moon struck the Earth,' said Isidor Rabi in 1934.[32] In Rome, the young Italian physicist Enrico Fermi pounced on the idea. His group used neutrons to breed artificial nuclei, but in doing so stumbled upon a new discovery: by first passing neutrons through a block of wax, they were slowed down, but nevertheless became more effective. It was the first example of a 'nuclear moderator'.

At Bohr's institute in Copenhagen, the papers describing the Rome results had been translated by Otto Frisch, who passed the news to his aunt. In 1934 Lise Meitner was still in Berlin, but was no longer working with Hahn. But she advised him that it was time to start looking at neutrons.[33] Hahn, now working with another nuclear chemist, Fritz Strassmann, took up Lise Meitner's suggestion and bombarded uranium with neutrons, uncovering in the process a profusion of highly unstable nuclei. Suspecting the appearance of new exotica, they redoubled their efforts. The opportunity to name a new nucleus or radioactive decay process was always good for scientific prestige. Other groups were now also busy. Big radioactive nuclei were expected to throw off small fragments, and this prejudiced experimenters. Lise Meitner, now following these developments from Sweden, could not understand their conclusions and wrote to Hahn, urging him to look more carefully.[34]

As Christmas 1938 approached, Hahn and Strassmann's careful chemistry now revealed that a single neutron could make a fat uranium nucleus split into two roughly equal parts.[35] Such radical nuclear transformation had never been seen before. The astonished Hahn, bereft of insight, wrote back to Lise Meitner: 'perhaps you can suggest some fantastic explanation'.[36] To keep his tradition of spending the holiday with his aunt, Frisch went to Sweden that year. Lise Meitner opened Hahn's letter just after Frisch had put on his skis. She set off after him, waving the letter. After she caught up with him, they stopped for him to read it, and began to glimpse what neutrons could do to nuclei. When a fat uranium nucleus absorbs another neutron, the delicate nuclear balance is broken. Instead of shedding mere fragments, the parent nucleus instead abandons hope and splits into two halves of comparable size. But there was something else. Lise Meitner realized that, taken together, the two daughter nuclei weighed slightly less than their parent. In 1909, when still a student, at a scientific meeting in Salzburg she had heard Einstein explain his equation $E=mc^2$. Both the scientist and his equation had left a strong impression on her. The velocity of light, c, is already a big number[37]: c^2 is gigantic. The tiny mismatches in nuclear mass resulting from uranium fragmentation should release huge amounts of energy. The seed of a new awareness was planted. Returning to Copenhagen, Frisch informed an astonished Bohr. 'How could we have missed that!' Bohr exclaimed, hitting his head with his hand.[38] In January 1939, Frisch and Meitner sent a joint letter to the science journal *Nature* in London, describing this new form of nuclear reaction, for which Frisch now coined the name 'fission', borrowing the term from cell biology. Bohr was about to leave for a trip to the United States, and Frisch showed him a draft of the letter before he left. Thus did the news of nuclear fission travel across the Atlantic.

In Berlin, Hahn and Strassman continued with their experiments, but with Lise Meitner no longer easily accessible, the interpretation of neutron developments moved to Bohr's group in Copenhagen. In residence there was George Placzek. Born in 1905 in Brno, in what is now the Czech Republic but was then part of the Austro-Hungarian Empire, he was the eldest son of a Jewish textile manufacturer.[39] After university studies in Vienna and Prague, he joined the peripatetic movement of young scientists in central Europe, beginning his research apprenticeship first at Utrecht, then in Leipzig with Debye and Heisenberg, before settling briefly in Rome where Fermi was investigating neutron irradiation. After Rome, in 1932 he moved to Bohr's Institute in Copenhagen, there to meet Frisch. When Frisch returned to Copenhagen after his Christmas 1938 epiphany, Placzek was on hand when the word 'fission' was first used in a nuclear context. Placzek advised Frisch to do an experiment to look for the nuclear effects that he and his aunt had predicted. Thus was Frisch the first to confirm their interpretation of fission, but he omitted to inform Niels Bohr, by now on the other side of the Atlantic.

Placzek firmly believed that travel broadened his mind. Frisch describes how the extrovert Placzek, despite having been a nomad for about a decade, was chronically incapable of organizing travel arrangements and packing his belongings.[40] After a brief residence in Kharkov, he had been offered a position at the new University of Jerusalem, at face value a great honour for a Jewish scientist. But it took the combined efforts of all his friends in Copenhagen to ensure that he got to the train station just in time to catch his train, which was fortunately one minute late. In Jerusalem, Placzek was supposed to learn Hebrew and give his lectures in that language. This did not deter a polyglot who had already mastered Czech, German, Danish, and English, and in Jerusalem he acquired some Arabic as well. Attempting to lecture on nuclear physics in Hebrew, he discovered that its vocabulary was deficient what it came to this branch of modern science. After that first year, his colleagues received a message from Jerusalem: 'Through with Jews forever'. Placzek soon reappeared at Copenhagen. He felt comfortable there: 'Why should Hitler invade Denmark? He can just phone can't he?' he observed.[41] But he soon felt that it was time to cross the Atlantic. Thus he was on hand when Niels Bohr arrived in the USA early in 1939.

As the news of nuclear fission crossed the Atlantic, Lise Meitner, isolated in neutral Sweden, became cut off from the physics that she had understood better than almost anyone else. In 1940, she briefly crossed back to Copenhagen to do experiments, but had to rush back to Sweden when the Germans invaded Denmark. Bohr told her that when she got back to Stockholm, she should send a telegram to London to tell them what had happened. This was to cause more confusion than reassurance.[42] An attempt to get her a job in Cambridge failed, as wartime Britain had initially placed radar as its top priority scientific research.[43] In 1945, the advancing Allied armies discovered the horror of the concentration camps. She was so shocked that she cried and could not sleep.[44] She tried to write to Hahn, but communications with Germany had broken down, although her letter survived: 'You all worked for Nazi Germany and did not even try passive resistance ... Millions of innocent human beings were allowed to be murdered without any kind of protest being offered.' After the war, Lise Meitner studied German philosophy and literature to try to understand what made them do what they did.

As the Allied armies advanced across Germany in 1945, Hahn was one of the scientists captured and interned by the British. Under house arrest, their every word was overheard. The eavesdroppers heard how Heisenberg, reading the newspaper,[45] informed an astonished Hahn that he had been awarded the back-dated 1944 Nobel Prize for Chemistry 'for his discovery of the fission of heavy nuclei'. The decisions of the Nobel committees are rarely easy: this time they were complicated by the war and its outcome. There was an immediate outcry—'Why not Lise Meitner?' She had been Hahn's long-term research partner and, with Otto Frisch, had explained what Hahn could not.

In 1945, when the atomic bombs were dropped on Japan, reporters pestered her. She told them little, and many of their reports were complete fabrication, with 'Fleeing Jewess' headlines making her a new icon for the bomb.[46] During a trip to the United States in 1946 to see her sisters, Lise Meitner was greeted as a heroine and showered with honours. A Hollywood version of her life story was a travesty of what really happened: she had not carried a bomb in her handbag.

After sitting, sometimes uncomfortably, in a series of niches in Stockholm, where she never really mastered Swedish, she moved to the Royal Technical University (*Kungliga Tekniska högskolan*).[47] In 1960, she left Sweden and retired to Cambridge, where her distinguished nephew was now a professor. Her brother Walter had moved to London in 1939, where his wife Lotte Meitner-Graf became a society photographer. Lise Meitner died in 1968—the same year as Otto Hahn, and at the same age, 89—and was buried in England. To the inscription on her tombstone, which reads 'A physicist who never lost her humanity.' could perhaps be added 'when all around were losing theirs.' Germany tried to make amends. In 1959, a new Hahn–Meitner Institute for Nuclear Research was established in West Berlin, and in 1982, a German team in Darmstadt discovered a new radioactive nucleus, number 109 in the Periodic Table. They called it meitnerium.

Before Hahn's award, the 1943 Nobel Chemistry prize had also been attributed for work on radioactivity: to the Hungarian émigré Georg von Hevesy for his development of radioactive tracers which pioneered the new science of nuclear medicine. He had also helped discover a missing element, hafnium, after the Latin name for Copenhagen, where he was then working. He moved on to Germany, but with Jewish ancestry overriding his newer Roman Catholicism, in 1935 left to rejoin Bohr's group in Copenhagen. There, with the outbreak of the Second World War, he dissolved the Nobel gold medals of Max von Laue and James Franck in acid to prevent them falling into the hands of the Nazis. The gold was successfully recovered after the war and recast into medals in Stockholm. When Denmark was taken over by the Nazis in 1943, von Hevesy moved to Stockholm.

Lise Meitner was not the only talented Austrian woman Jewish physicist to become a victim of the Nazi edicts. Marietta Blau, born in Vienna in 1894, studied there and worked in a series of research positions funded by the Austrian Academy of Sciences, where she helped to develop new photographic techniques for studying subnuclear interactions. In 1938 she left Austria, moving first to Oslo, then Mexico, and the USA, before returning to Austria in 1960.

Italy

While Hahn and Strassmann had been polishing their results on what would soon become known as nuclear fission, in Rome, Enrico Fermi was packing his bags for a trip to Stockholm. He was going to receive the 1938 Nobel Prize for Physics. It marked what had been a new renaissance in Italian science. Long before Italy existed as a nation, Italian artists and humanists had been the spearhead of the European renaissance, whose message for science was embodied by Galileo. But in the twentieth century, Italian science had dropped behind. Fermi was designated to lead the chase. Rome was on its way to becoming a new centre of science. Born there in 1901, Fermi had been very close to his older brother Giulio, who died in 1915 after what appeared to be a trivial throat infection. Stunned, Fermi immersed himself in science, and what had begun as a childhood pastime became a lifelong career.

In 1927, he came to the attention of the University of Zurich, looking for a successor to Schrödinger after his departure for Berlin. But Fermi's interest in the move palled when he married Laura Capon, a Rome science student from a Jewish family. The Italian authorities had encouraged Zurich to recruit him, and were irritated when Fermi finally turned down the offer. They had seen the move as a way of enhancing Italy's scientific reputation.[48] However, with Fermi now firmly anchored in Rome at the head of a talented young group, his laboratory soon became a new port of call for European nuclear physicists. Such a reputation needed adequate support, and in 1937 Fermi pushed for a substantial increase in funding. After his patron Orso Corbino died in 1937, this request was refused, and Fermi began to think about a move to the United States. Even before his key nuclear discoveries, Fermi had been invited to cross the Atlantic, where he introduced himself to English by reading novels by Jack London.

Soon after his request for additional funding was turned down, Fermi became entangled in Mussolini's race laws. Although not a Jew, he was married to one, and now had children, so his family was particularly vulnerable. Although Jews were only a small minority in Italy, this was not the case in Fermi's group in Rome. The 1938 racial ordinance was the last straw (*la gocchia che fece traboccare il vaso*).[49] The group quickly dispersed. The great days of Italian nuclear physics and the country's second scientific renaissance lasted less than a decade.[50]

Fermi's colleague Franco Rasetti was also not Jewish, but was perturbed enough by his experiences in Rome to leave for Canada in 1939. With no wish to contribute to military applications, he switched to other branches of science. Giulio Racah, born in Florence in 1909, left Italy in 1939 to become (as Yoel Racah) Professor of Theoretical Physics at the Hebrew University of Jerusalem, making major contributions to the mathematical foundations of atomic physics. His career paralleled that of his cousin Ugo Fano, born in Turin in 1912, who also worked with Fermi, and later with Heisenberg in Leipzig, before leaving for the USA in 1939. There he studied molecular biology and the biological effects of radiation. The Bemporad brothers Azeglio, an astronomer, born in Siena in 1875, and Giulio, a mathematician, born in Florence in 1888, had to leave their respective posts in Naples and Turin in 1938. Another who left was mathematician Eugenio Fubini (1913–97), who went on to become a vice-president of IBM. Nella Mortara, born in Pisa in 1893, had also worked in Rome, assisting with Fermi's radioactivity studies. She fled to Brazil in 1939 with her elder brother Giorgio, a statistician working in Milan, but managed to return to Italy in 1941. After the war, she became a professor at Rome.

Tullio Levi-Civita (1873–1941) had helped develop the tensor calculus which became an indispensable tool for Einstein's theory of general relativity. After the race laws barred him from his prestigious professorial chair in Rome, he died in obscurity. Rita Levi-Montalcini, born in Turin in 1909, embarked on research work with Giuseppe Levi, another Jewish scientist, just before the Second World War, but the pair carried on working clandestinely. After the war, she moved to Washington University, St. Louis, and in 1986 shared the Nobel Prize for Medicine for her work on the neurology of embryos. In 2001, the Italian President appointed her Senator for Life. She was just one Nobel laureate student of Giuseppe Levi: another was Salvatore Luria, who made his way to the USA in 1940, where he was helped by other Italian emigrants and went on to do research with Max Delbrück, with whom (and Alfred Hershey) he shared the 1969 Nobel prize for Medicine for their work on the phage viruses that infect bacteria.

Also a student of Giuseppe Levi was Renato Dulbecco, who after the war moved to the USA with the help of Luria and Levi-Montalcini, and went on to earn the 1965 Medicine Prize for his work on the transcriptase enzyme which converts RNA to DNA. Before the emergence of Fermi on the nuclear scene, Bruno Rossi, born in Venice in 1903, had made important contributions to nuclear physics through his studies of cosmic rays. As an Italian Jew, Rossi left his professorial post at Padua in 1939, and after transiting via Copenhagen and Manchester, where he briefly worked with the benevolent Blackett,[51] moved to the University of Chicago, before Bethe summoned him to join the Atomic Bomb work at Los Alamos. The invitation demanded a lot of introspection. Later, he related, 'I clearly remember my feelings when I decided to go to Los Alamos. I was hoping that our work would prove that the fission bomb was not feasible. However, I had also reached the conclusion that if, on the contrary, the bomb was feasible we must make sure, at all costs, that Hitler did not have it before we did.'[52] He was soon joined by another Italian émigré, Emilio Segrè, who had already made important discoveries at Palermo. Sergio de Benedetti, born in Florence in 1912 and who had done cosmic-ray research with Rossi, left Italy to work with the Joliot–Curies in France before moving to the USA.

Fermi's Nobel award acknowledged his work with neutrons. Normally, Nobel prizewinners are startled by early-morning telephone calls from Stockholm, but this time Niels Bohr had leaked the news to Fermi. With Hitler having made problems for German Nobel laureates, Bohr did not want Fermi to fall foul of another dictatorship. Fermi and his wife saw that a trip to Stockholm would be a convenient way of leaving Italy: they would arrive in New York with Nobel prize money to assure their future. After the Stockholm festivities, Fermi had passed through Bohr's Institute in Copenhagen en route to Southampton for their Atlantic crossing. They must have worried together about developments in Europe, as well as discussing neutron science, but at that time, Bohr and Frisch had no startling news to reveal.

Bohr arrived in New York on 16 January 1939, two weeks after the Fermis. Bohr had promised Frisch that he would keep his mouth shut about the post-Christmas fission news until Frisch and Meitner's paper had appeared in print. But Bohr had crossed the Atlantic with another scientist, the Belgian Léon Rosenfeld, and during the crossing they did nothing else but talk about nuclear fission. Bohr omitted to tell Rosenfeld about his promise to Frisch. On arrival in New York, Bohr was greeted by Fermi, while Rosenfeld continued to Princeton, where he promptly gave a talk about the new phenomenon of nuclear fission. In the audience was Isidor Rabi, who brought the news to Columbia University, where Fermi was getting ready to do the experiment that Frisch in Copenhagen had already done.[53]

Before Joliot was invited to Stockholm in 1935, there to publicize his chain reaction idea, another scientist had realized. After helping to launch the effort in London to support the new wave of refugees, in 1933 the perceptive Leo Szilard saw clearly the possibility and implications of a nuclear chain reaction. Ernest Rutherford, the ageing father of nuclear physics, had famously dismissed the possibility of generating energy from nuclear reactions as 'moonshine'.[54] But Rutherford had passed his prime and did not fully realize the potential of neutrons. So confident was Szilard about his scheme that he took the precaution of taking out a patent, lodging it with the British Admiralty for safety. Without an invitation, he strode into Oxford University, proposing to install himself in the laboratory. Away from science, Szilard's sensitive political predictor was still active. He had told Michael Polanyi in Manchester that he would stay in Britain until one year *before* war broke out in Europe, and then move to the USA. Polanyi wondered how his Hungarian colleague could time the future so accurately. Szilard turned up in the USA in January 1938, nine months plus one year before the outbreak of war. Apparently his predictor needed some fine tuning. When Niels Bohr's ship docked in New York one year later, three exiled Europeans—Fermi, Placzek, and Szilard—became the leaders of a group which picked up the baton of nuclear fission and ran with it.

NOTES

1. Szanton, p.199.
2. Sime, p.3.
3. Sutton, C., *Spaceship Neutrino*, Cambridge University Press, Cambridge, 1992, p.11, attributed to Meitner.
4. Hahn, O., *Mein Leben*, Brückmann, München, 1968.
5. Comment by Max Perutz, quoted in Bernstein, *Plutonium*, p.37.
6. Sime, p.150.
7. Hargittai, p.48.
8. Rhodes MAB, p.80.
9. James, p.147.
10. Bernstein, *Plutonium*, p.45.
11. Rhodes MAB, p.235, and Frisch, O., In Lise Meitner, *Biographical Memoirs of Fellows of the Royal Society*, 1970, Vol 16, pp.405–20.
12. Sime, p.184., quoted in Dahl, P., *Heavy Water and the Wartime Race for Nuclear Energy*, Institute of Physics Publishing, Bristol, 1999, p.66.
13. Sime, p.189.
14. Sime, p.195.
15. Frisch, O., *Biographical Memoirs of Fellows of the Royal Society*, 1970, Vol 16, p.405.
16. Sime, p.204.

17. Rhodes MAB, p.236, attributed to an article by George Axelsson in the *Saturday Evening Post*, January 5, 1946.
18. Herwig Schopper, private communication 7/3/2011.
19. Frisch, O., SPSL, Box 335.
20. Vogel, R., *Ein Stempel hat gefehlt*, Droemer Knaur, Munich, 1977, p.44. From 1933 to 1939, the total emigration tax paid increased from 1 million to 342 million RM. (Strauss, H. and Kampe, N. (eds), *Jewish Emigration from Germany 1933-42*, Saur, 1992, Vol 6.)
21. *The* Adventures of F G Houtermans, In Amaldi, Battimelli, and Paolini, p.19.
22. Cahen, F.M., *Men Against Hitler*, Jarrolds, London, 1939.
23. Sime, p.205.
24. Sime, p.207.
25. Fermi, L., quoted in Dahl, P., *Heavy Water*, 1999 (see note 12), p.74.
26. Frisch, p.3.
27. It was in the process of changing, and would only become complete in 1938.
28. Frisch, p.34.
29. Also at Blackett's haven at Birkbeck at the time was Werner Ehrenberg, born in Berlin in 1901. Ehrenberg later moved to Kharkov, before finding more permanent employment at the EMI electronics concern near London. In 1940, he was interned with other German expatriates on the Isle of Man, but was released on medical grounds. Blackett, then at the Royal Aircraft Establishment, Farnborough, had pointed out that Ehrenberg was anti-Nazi and a cripple.
30. Joliot, 1935 Nobel Prize lecture.
31. Joliot and Curie had almost got there first.
32. Quoted in Bernstein, *Plutonium*, p.23.
33. Meitner, L., IAEA Bulletin, December 1962, quoted in Rhodes MAB, p.234.
34. Frisch, see note 15.
35. The idea had been aired several years earlier by the husband and wife team of Walter and Ida Noddack at Freiburg, but nobody had taken their suggestion seriously. Even Hahn had thought it 'ridiculous', Jungk, p.65.
36. Rhodes MAB, p.253.
37. Some 300,000 kilometres per second.
38. Jungk, p.73.
39. Fischer, J., *George Placzek – an unsung hero of physics*, CERN Courier, September 2005, p.25.
40. Frisch, p.81.
41. Rhodes MAB, p.243.

42. See Chapter 10.
43. Dahl, p.211 and Gowing, p.52.
44. Cornwell, p.411.
45. Bernstein, *Plutonium*, p.38, citing the diary of Erich Bagger.
46. Sime, p.315.
47. Herwig Schopper, private communication 7/3/2011.
48. Orlando, L., Physics in the 1930s, Jewish physicists' contribution to the realization of the new tasks of physics in Italy, In *Historical Studies in the Physical and Biological Sciences*, Vol 29, 1998, p.151.
49. Finzi, R., *Le Leggi Razzialai e l'Universita Italia*, Editori Riuniti, Roma, 1997.
50. For a full list, see Zimmerman, J., *Jews in Italy under Fascist and Nazi rule*, 1922–45, Cambridge University Press, Cambridge, 2005.
51. Thereby turning down an offer from Weizmann to work in Jerusalem (note 52, p.75).
52. Bonolis, L., Bruno Rossi and the Racial Laws of Fascist Italy, *Physics in Perspective*, 2011, 13, p.82, citing Rossi, B., *Moments in the Life of a Scientist*, Cambridge University Press, Cambridge, 1990, p.68.
53. Dahl, p.198.
54. Rhodes MAB, p.27.

9

Gathering nuclear fuel

When the First World War broke out in 1914, Britons who had not been able to get out of Germany in time were transferred to a hastily prepared camp on a racecourse near Berlin. It was called Ruhleben, literally 'quiet life'. Among them were some footballers who had been playing with or coaching German teams, some sailors from ships berthed in German ports, a few newspaper correspondents, and a scientist, James Chadwick (1891–1974). After studying at Manchester and being influenced there by Ernest Rutherford, in 1913 Chadwick had been awarded a scholarship. He wanted to stay at Manchester and continue working with Rutherford, but the scholarship stipulated that he had to use it somewhere else.

In German, *Geiger* means 'violinist', but elsewhere in the world, the word is usually associated with the clicking counter that measures radioactivity. Its inventor, Hans Geiger (1882–1945), studied physics at Erlangen, then went to work with Rutherford in Manchester, helping with his classic experiments. There they pinpointed the tiny nucleus hidden at the heart of the atom. In 1912, Geiger returned to Germany to take up a post at Berlin's *Physikalisch Technische Reichsanstalt*. Following the Manchester connection, Chadwick chose to use his scholarship to go and work with Geiger.

In 1914, Geiger, a reserve officer, had to go into the army. Chadwick, looking at ways of returning to Britain via Switzerland, was overheard in a Berlin street making inappropriate remarks about the course of the war,[1] and was sent to the Ruhleben racecourse stables. Each internee had his own way of passing the time. There were lots of football matches. Chadwick preferred to do scientific experiments using German toothpaste as a faint source of radioactivity.[2] Also at Ruhleben was Charles Ellis (1895–1980), an army cadet who had been holidaying in Berlin when war was declared. In his four years in the camp, Ellis learned a lot of science from Chadwick, and later became a key member of Rutherford's group at Cambridge.

Another Ruhleben internee was Geoffrey Pyke (1893–1948), who had been a London newspaper correspondent in Berlin. Pyke came from a family of religious Jews, who made the unfortunate mistake of sending him to an élite British school while still dressed as an orthodox Jew, an embarrassment he had to live with for several years. More conventionally dressed, he was arrested as a spy in Berlin, Pyke was initially kept in solitary confinement before being transferred to Ruhleben, from where he and a colleague escaped to Holland. Pyke later became known for his imaginative but idiosyncratic ideas, such as his Second World War scheme for making aircraft carriers from 'pykrete'—ice reinforced with sawdust. It was never implemented.

After the war, a malnourished Chadwick returned to England and rejoined Rutherford in Manchester. Dutifully, he wrote a full report to his scholarship funding agency on his improvised experiments at Ruhleben. When Rutherford moved to the Cavendish Laboratory in Cambridge the following year, Chadwick went with him. Rutherford was now convinced that the atomic nucleus had to contain much more than just protons. 'I think we shall have to make a real search for the neutron', declared Chadwick in 1924.[3] This 'real search' would take some time. It could have ended early in 1932 when the Joliot–Curies in Paris reported something new, but they misinterpreted what they had seen. However Chadwick understood, and in a ten-day spurt finally achieved what he and the rest of the Rutherford team had been trying to do for more than a decade. For this, Chadwick earned the Nobel Prize for Physics in 1935. In Stockholm he met the Joliot–Curies, who were there to receive the Chemistry award. Such was the fruitfulness of radioactivity in the 1930s.

With the missing component of the nucleus discovered, the objective switched to what these neutrons could do. As Rutherford's dominance of nuclear physics began to wane, Enrico Fermi in Rome moved to the front of the pack. But a quiet neutron coda was played in Cambridge. In 1933, a young Jewish student called Maurice Goldhaber had arrived, eager to work with Rutherford and his distinguished colleagues. Born in what is now Lviv in the Ukraine in 1911, Goldhaber had studied physics in Berlin, where he saw and heard Einstein. But the events of 1933 interrupted his studies, and he decided to move to Cambridge.[4] There, with Chadwick, he carefully weighed the neutron. It was about one part in a thousand heavier than the proton. In nuclear accounting, where discrepancies get multiplied by the square of the velocity of light, even such a tiny number was very significant.

(The Goldhabers were an itinerant tribe. After Maurice Goldhaber left for Cambridge, his family, with his younger brother Gerson, left Germany for Egypt. Gerson later studied physics at the Hebrew University of Jerusalem, where he met his wife Shulamit, another young Jewish emigrant, born in Vienna. Both went on to have distinguished careers at the University of California at Berkeley. After moving to the University of Illinois in 1938, Maurice Goldhaber married the talented Gertrude Scharff, born in Mannheim in 1911, whom he had first met during their studies in Berlin. After completing her doctorate at Munich in 1935 with Sommerfeld and Gerlach, Gertrude Scharff had moved to London.)

Neutrons and protons are the twin denizens of the nuclear world. As with biological twins, there are subtle signs and personality traits which differentiate them. For the nuclear pair, the most obvious is their electric charge: the protons supply the positive charge to balance the negativity of orbital atomic electrons. Neutrons, as their name implies, have little atomic electrical role: their business is essential nuclear. Then there is the discrepancy in their mass, measured by Chadwick and Goldhaber. Nuclear theory, a mathematical formalism to describe this duality of neutrons and protons, was provided by another quantum exile. As though emulating the proton and the neutron, he himself appeared to be inseparable from a companion. Thirteen months separated the births of Eugene Wigner (1902) and John von Neumann (1903) into Jewish families in Budapest. While their birthplace and their respective talents made their schooling, studies, and scientific careers overlap and interconnect, their personalities were very different. But this disparity somehow complemented their individual scientific contributions.

Wigner's father was the manager of a flourishing leather business, and he wanted it to continue flourishing under his son. For a time it did, but it was clearly a waste of some remarkable intellectual abilities. Wigner's name is now written all over physical science: there are equations, effects, theorems, forces, and structures named after him: for hundreds of years to come, those who know nothing about the man will still know of him. Wigner and von Neumann first met at Budapest's Lutheran Gymnasium. Secondary school students traditionally advance in one-year steps, so even the one-year age difference between Wigner and von Neumann could have separated them. But a gifted mathematics teacher, Lazlo Ratz, soon recognized their talent, and they were thrust together, where the younger von Neumann set the pace. As well as learning from his mathematics teacher, Wigner also learnt from von Neumann.

When the 1919 communist revolution in post-war Hungary aimed to nationalize private enterprise, small businesses like that of the Wigner family were ready targets, so the family laboriously uprooted themselves and moved to neighbouring Austria. Their stay was as short-lived as the post-war Hungarian communist regime. Within a few months, they were back in Budapest. Wigner's father, with an eye to his leather business, pointed his son towards a future in chemical engineering, with initial studies in Berlin. But once there, like other talented but impressionable students, Wigner was attracted by its heavyweight physics school, with Einstein at its lofty pinnacle. Von Neumann also contrived to study chemical engineering in Berlin, and was likewise impressed by the stars in its physics and mathematics firmament. The duo were soon noticed by another Hungarian, Michael Polanyi.[5]

For his doctorate, Wigner worked on molecular forces under Polanyi's guidance, but became intrigued by the structure of sulphur crystals.[6] This introduced him to another field of mathematics—symmetry—which he would revisit later. Wigner, with a strong sense of family responsibility, returned to Budapest in 1925 to work in his father's factory. But his scientific abilities had not been forgotten in Berlin. Soon he received an unexpected offer, engineered by Polanyi, of a research position in Berlin's prestigious crystallography group, using X-rays to reveal the atomic structure of crystals. However, this was a dilemma for the Wigner family: both father and son were committed to the success of their business, but they recognized the importance of making the most of scientific ability. After much soul-searching, Wigner returned to Berlin. Entering a new field of research at the deep end is not easy. Wigner was soon confused, and asked von Neumann for help. His recommendation of a totally different mathematical approach was another turning point in Wigner's scientific career.

What distinguishes crystals from powders is their characteristic shape, their symmetry, like the cubes of sodium chloride, common salt. For mathematicians, a symmetry is the trademark of an underlying mathematical structure, a 'group', and the technique recommended by von Neumann was that of group theory. Regular shapes look the same when rotated through an angle, three times 120° for an equilateral triangle, four times 90° for a square. Each set of such rotations is the symmetry group of a geometrical shape. As well as being handy for describing crystals, Wigner realized that these techniques also provided a powerful tool for the new quantum theory. It was a way of preserving valuable intuition, which was otherwise in danger of being lost in the unfamiliarity of the quantum world.[7]

Wigner's mathematical prowess soon became noticed in Göttingen, and in 1927 he moved to work under the eminence of David Hilbert, and in the company of imaginative young scientists such as Walter Heitler and Victor Weisskopf. Wigner's commitment to the mathematics of quantum mechanics deepened. His Hungarian colleague Leo Szilard, then in Berlin, wisely suggested that he should write up his group-theory findings so that other scientists could benefit. However, before the book was completed, Hermann Weyl in Zurich produced one on the same subject, which became widely acknowledged as a classic by those who could understand it. The problem was that most people found it incomprehensible. This was rectified in Wigner's book, which went on to be even more influential.

In 1928 Wigner returned to Berlin, where he began work with Polanyi on the dynamics of chemical reactions, but his concentration was broken by an unexpected invitation from Princeton. The career paths of von Neumann and Wigner had overlapped again. Princeton had invited von Neumann to join its mathematics department. Although he was keen to move across the Atlantic, he was reluctant to leave Europe definitively, and suggested instead that Princeton hire him and his old school pal together on a 50:50 basis. So the pair moved in 1930. While Princeton was getting a two-geniuses-for-the-price-of-one bargain, the different personalities of the two Hungarian immigrants were reflected in their respective adaptations to life in the USA. The extrovert von Neumann took to American life like the proverbial duck to water, but the reserved Wigner quickly became isolated. His Central European politeness became a campus legend, where his colleagues learned that he would never allow himself to pass through a door before anyone else. Developments in Germany led Wigner to believe that 'Europe was a sinking ship',[8] and spurred him on to take US nationality in 1937. Hungary, where Wigner's parents still lived, was not far from Germany, and anti-Semitism was endemic. In spite of their formal conversion to Christianity, his parents were still Jews by Nazi standards. In 1939,

Wigner paid for them to emigrate to the USA. Neither spoke any English, and they became very lonely, despite their son's attention. It was a difficult transition from being a factory manager to an apparently illiterate nobody.

Wigner helped develop techniques for handling and interpreting nuclear data, charting the forces which gripped protons and neutrons. However, his accomplishments in what was nominally a half-time post at Princeton had stirred up some professional jealousy, and in 1936 he was 'shocked' to be fired from Princeton. Feeling that he had been overshadowed by the intellectual fireworks of von Neumann, he transferred to Wisconsin, where he married Amelia Frank, a faculty scientist. The bachelor who had been so lonely at Princeton enjoyed marital bliss, and was deeply involved in his work. It could get no better. But as though fate refused to allow him to be contented, his wife died of cancer just a few months after their marriage. Realizing their mistake, Princeton now invited Wigner back, this time with a tenured position. Although Wigner still had misgivings about Princeton, he wanted to leave the scene of his tragic marriage. Princeton is also closer than Wisconsin to New York. Thus, another European emigrant neutron expert was on hand when Fermi and Bohr arrived from Europe in close succession.

'A new and important source of energy'

On his arrival in New York in December 1938, Fermi registered at a hotel opposite Columbia University. Immediately, up popped Szilard, who had already begun neutron investigations at the university. Neutron physics had suddenly became very popular. Two weeks later, Szilard and Fermi went to the dockside when the ship carrying Niels Bohr berthed and were immediately vitalized by his news of the discovery of fission. The final piece of Szilard's nuclear jigsaw fell into place, and as a precaution he cabled the British Navy, which still held his patent on nuclear energy. Szilard the predictor had become Szilard the pacemaker. Experiments at Columbia—his own and those of his new American colleagues—were suggesting that if enough uranium could be amassed, the predicted nuclear chain reaction looked likely. Szilard telephoned his colleagues: 'I have found the neutrons'.[9] But far from being triumphant, Szilard was instead gloomy. He had seen the future again. If enough uranium could be amassed, it could release enough energy to affect the course of history. All of the European experts were at a scientific meeting in Washington in January 1939, and the resultant speculation was picked up by the alert *New York Times*. Nuclear fissions (the *Times* did not use this word, preferring 'electrical forces many hundred times more powerful than the force of gravity')[10] could unleash a new source of energy. A US wire service's telegraphese was more blatant—'Is world on brink of releasing atomic power?'.[11]

Earlier, in Rome, Fermi had discovered the importance of neutron energy: intruding neutrons can appear temperamental, enigmatically becoming more effective if they are first calmed down. Frisch in Copenhagen was pushing forward with fission. After looking at uranium, he had gone on to investigate another heavy metal residing at the far end of the Periodic Table of elements. Hahn in Berlin had investigated thorium, but it was Frisch who found that its fission looked very different from that of uranium. Slow neutrons did not do the job. This time Frisch could not explain why. The answer was to come from Bohr, now in temporary residence in Princeton, where George Placzek had also arrived. Placzek pointed out to him the puzzling irregularities in neutron behaviour with heavy nuclei. Bohr drew some curves on the blackboard and thumped his large head with his hand.[12]

There is more than one kind of uranium. Most atoms of natural uranium have nuclei with 238 particles—92 protons and a swarm of 146 neutrons. But less than one per cent of natural uranium is of a different kind, with three neutrons less, containing 235 nuclear particles. This slightly reduced neutron complement is enough to make the rare uranium isotope behave very differently—it can be split by any neutron, slow, as well as fast. The traditional response of any scientist after making such a discovery is to write it up and submit it for publication. Just as he had done after Frisch had explained what he and his aunt had understood during their Swedish Christmas, so Bohr did one month later, this time underlining the special role of uranium 235.

The increased fission yields of uranium 235 made it the most combustible nuclear material then known. Most people wrote it off as a scientific oddity, a nuclear pollutant, but others mused that, in theory, it could be used to make a nuclear bomb. The vigilant *New York Times,* attending the Spring 1939 meeting of the American Physical Society, reported 'Niels Bohr of Copenhagen, a colleague of Albert Einstein at the Institute for Advanced Study, Princeton, declared that bombardment of a small amount of the isotope uranium 235 with slow neutron particles . . . would start a 'chain reaction' or nuclear explosion sufficiently great to blow up a laboratory and the surrounding country for many miles'.[13] To underline the significance of the news, the name of Einstein was being linked to scientific developments in which he was only a spectator.

In spite of this media enthusiasm, not all scientists were convinced. And there remained the difficulty of isolating uranium 235: it did not grow on geological trees. It was a nuclear contaminant, a tiny trace hidden inside natural uranium. Moreover, it could not be separated from its host metal by dissolving it in acid or by any other chemical means. For chemistry, all uranium nuclei look the same. Meanwhile Fermi and Szilard set out to discover what would happen if ordinary uranium were piled up. They knew that a small amount would just sit there, its natural radioactivity just a low nuclear hum, its neutrons escaping. But above some critical size and encased in a blanket of carbon graphite, more neutrons would stay inside, generating additional neutrons by fission. Then the uranium–graphite pile would start to produce energy all by itself. Would the assembled mass slowly heat up, or would it explode? How much uranium would be needed? Some said thirteen tonnes, others more than three times this amount. Fermi and Szilard carefully began assembling uranium and graphite, brick by brick, keeping a careful eye on their neutron counter.

Science transcends national frontiers: what goes on in an American nucleus also happens in a European one. Whenever they find something, scientists respond by publishing it in a scientific journal. The publication date of the paper reporting a discovery is the traditional way to claim credit. Szilard knew what was going on in Europe. If Hitler wanted his Reich to cover the world, uranium fission could be a weapon to help him do it. There was nothing Szilard could do about science in Berlin, but at least he could try to stop it benefiting from the advances made by others. With developments following in quick succession, scientists were eager to stake their discovery claims. To control the spread of this nuclear know-how, Szilard wanted to muffle as many of these voices as possible. At first this was difficult, and scientists in Berlin avidly read uranium research papers in their air-mailed copies of *The Physical Review*. Eventually, Szilard convinced his colleagues, and the wave of neutron-related developments published in American physics journals slowed to a trickle. In Berlin, scientists who had been following the developments noticed,[14] and second-guessed that something had happened across the Atlantic.

Fermi's objective was to increase his accumulated amount of uranium gradually, coaxing his neutrons to obtain a smooth release of energy before there was any danger of an explosion. There had been open talk of uranium bombs at physics meetings, but still no one knew how much uranium would be required. Fermi's team was piling up uranium in New York but no sign of activity had yet appeared. Would a bomber even be able to take off with the amount of uranium needed? The new American B-17 Flying Fortress could carry a long-range bomb load of about two tonnes, far less than the amount of material already in Fermi's hands at Columbia. German bombers then flying carried similar loads.[15] And how could any uranium bomb be kept safe in transit and suddenly made to detonate when it reached its target? Any explosion looked increasingly distant: instead, nuclear power looked more likely to be a source of energy—a 'uranium burner' whose flames needed no oxygen.[16]

Fermi's laboratory in New York was not the only scene of uranium research in 1940. In Paris, Frédéric Joliot and Irène Curie still had a tradition to maintain. A fervent communist, Joliot had many contacts in the Soviet Union, and in January 1939 told his scientific colleagues in Leningrad about the discovery of nuclear fission. At about the same time as Niels Bohr's ship docked in New York, the same nuclear message also arrived in the Soviet Union.[17] French neutron studies had been reinforced by the arrival of two Jewish immigrants from further east. Hans von Halban was descended from Polish Jews who had settled in Vienna. His grandfather had achieved high rank in the Austro-Hungarian hierarchy, earning his descendants the right to use the honorific 'von'. His father was a chemist, and transferred the family to Zurich, where Hans completed his physics education, spending a year with Bohr in Copenhagen before moving on to Paris. His colleague there was Lew Kowarski, born in St Petersburg in 1907, from where his father fled to Vilnius during the Russian revolution, also having to negotiate the crumbling lines of the German eastern offensive. Kowarski inherited considerable musical ability from his mother, a singer, but he was a huge man and his stubby fingers were too big for a concert piano keyboard. After studies in Brussels and Lyon, he joined the Joliot–Curie team.

In the USA, Fermi and Szilard were using graphite to control neutrons in uranium. The Paris laboratory had instead selected 'heavy water', where the normal hydrogen of H_2O is replaced by deuterium, an isotope of hydrogen in which the lone nuclear proton is paired with a neutron. In ordinary water, about one 'hydrogen' nucleus in every few thousand is in fact deuterium, and this concentration can be increased by repeated distillation or electrolytic separation. Paris had cornered the world market for the exotic liquid, amassing 185 kilograms. When the German army invaded France in 1940, the entire laboratory could not be transported, but Halban (he soon dropped the 'von') and Kowarski first moved south to Clermont-Ferrand, where their precious heavy water was temporarily stored in prison cells while they made plans to export it to England. In a saga redolent of a James Bond adventure, the heavy water was taken aboard a British ship at Bordeaux under the supervision of the Earl of Suffolk, a flamboyant pistol-wielding British 'scientific attaché', looking more like a pirate than a diplomat. The Earl supervised the loading of the heavy water canisters and other valuable cargo, including industrial diamonds worth millions of pounds, onto wooden rafts to ensure they would not sink if the ship were torpedoed. Another vessel which left Bordeaux at the same time was sunk by a torpedo, allowing Joliot in Paris to leak a story that the precious heavy water had been lost. On arrival in England, the heavy water was transferred for safe keeping first to London's high-security jail at Wormwood Scrubs, and then to the more salubrious surroundings of the library at Windsor Castle.[18]

But even before the French heavy water embarked for London, Szilard's predictor had been buzzing alarmingly. The exact future of nuclear fission was still unclear, but for Szilard the dangers were real enough. It was time to start making political moves. He felt that he and the other European nuclear immigrants were the best qualified to beat the Nazis in any race to develop a bomb. They had the nuclear know-how, and they had also seen the Nazi threat at first hand. The Americans, on the other hand, were simply too far removed from such danger. The US military were approached, but were not convinced by presentations from foreigners speaking accented English. Fermi, with a Nobel Prize and in principle the spokesman, emitted only science-speak which his high-ranking audience could not follow. Szilard saw how his old friend Albert Einstein, still the most visible scientist in the world, could help.

The raw material for fission fuel was uranium ore, for which one of the world's major suppliers was the Belgian Congo. Einstein had lived briefly in Belgium before moving to the USA and had met Queen Elizabeth, now the mother of the nation's new monarch. Szilard's initial plan was to ask Einstein to get her to put pressure on the Belgian authorities, so that uranium ore from Belgian Africa would not make its way into German hands. Szilard had no car—he did not know how to drive—and to get a ride called on his Hungarian colleague Eugene Wigner, also at Princeton. However it was now July, and Einstein had left for his summer residence, so the pair had to go to the Long Island coast. Einstein, rarely smartly dressed, was even more casually attired than usual. He listened carefully to their plea. He had not himself realized the possibility of a nuclear chain reaction, neither had he read about it. However with his political awareness, sharpened by his experience in Germany, he promised that he would help in whatever way he could. He dictated, in German, the first draft of a letter that would eventually influence history.

Szilard and Wigner mulled over their plan. Perhaps Belgium should not be top of their agenda. Their Einstein card could be played to better advantage. They were now all in the USA, so rather than trying to influence developments themselves, maybe they should try to get the US authorities involved. Belgium would be more likely to listen to a plea from the US State Department than from some amateur scheme, even if it had Einstein's backing. Realizing that there was more political homework to be done, Szilard began networking. His connections told him that the target should be the White House, not the State Department. Einstein should write to President Roosevelt, alerting him to the possibility of nuclear weapons. The letter, signed by Einstein, was forwarded to the President in August 1939. It began, 'Some recent work by E. Fermi and L. Szilard ... leads me to expect that the element uranium may be turned into a new and important source of energy'. It went on to describe the construction of bombs of a 'new type' which 'might prove too heavy for transportation by air'.[19]

Several months before, Paul Harteck, a German scientist and an advisor to the German Army Ordnance Office, had warned of 'developments in nuclear physics, which in our opinion will probably make it possible to produce an explosive many orders of magnitude more powerful than the conventional ones. ... That country which first uses it will have an unsurpassable advantage over the others'.[20] Harteck, born in Vienna in 1902, had studied and worked mainly in Berlin before spending 1933 in Rutherford's group in Cambridge.

Thus, Germany had already latched onto the potential of nuclear fission. Szilard and the other emigrants in the USA did not know this, but would not have been surprised. Their stated goal was to develop nuclear fission into a weapon before the Germans did. After Harteck's warning, the Germans in fact got off the mark first, using a heavy water assembly along the same lines as Joliot in Paris. But Harteck was not as influential as Einstein. The German Army Ordnance Office had other priorities, and was remote from Hitler.

In September 1939, Germany invaded Poland and Europe was at war. Nuclear fission quickly lost its place in the list of politico-military priorities. President Roosevelt, the German Army Ordnance Office, and indeed almost everyone else had more pressing problems. Wigner said 'getting the US government to see the value of fission was like swimming in syrup. We learned that governments do not like fantastic new projects. And no scientist could honestly say that an atomic bomb was sure to work'.[21] Roosevelt finally saw the Einstein letter on 11 October, and ordered the creation of a new effort to investigate uranium 'so that the Nazis don't blow us up'.[22] Some money was provided to complement the meagre resources of university research departments. Einstein had helped Szilard to achieve his objective. Having done so, he played no active role in all that was to come. And nobody was quite clear what to do about bombs that 'might prove too heavy for transportation by air'. But the quantum exiles nevertheless had a clear idea of what was at stake. One of them, Victor Weisskopf, wrote of 'the fear of the Nazis beating us to it. … We were all acutely aware that the whole of the civilized world was under attack by a force of the greatest evil'.[23]

When Szilard went to see Einstein a second time, carrying the latest draft of the letter to the President, he had not been able to find Wigner at short notice, so his driver had been another Hungarian physics émigré, Edward Teller.[24] It would be the first time that Teller and Einstein had met, although many years before, Teller had gone to listen to Einstein speak about science in Germany and had been disappointed not to have understood.[25] Gifted children have to learn how to manage their brilliance: undiluted precocity can wear down parents, antagonize teachers, and annoy fellow-students. It is not good to upstage the whole classroom. Teller often had, and had not been popular. In the post-1918 trauma in Hungary, troops had been billeted in the Teller house. These country soldiers were not house-trained, and the young Teller was terrified. Budapest was not a comfortable place to be: lynch mobs roamed the streets, and thousands were killed. Teller's father told Edward sternly that he would have to emigrate when he was old enough to manage alone. The opportunity came in 1926, and he went to study in Karlsruhe. Two years later, in Munich, Teller slipped as he got off a tramcar and his right foot was severed under the wheels. He recalled looking at his boot and hoping that it was not damaged. That was before the pain started.[26] For the rest of his life, Teller had to wear a prosthetic heavy black boot, and walked with a limp. A physical scar had been added to the psychological damage of his Budapest childhood. Close to, Teller's face was dominated by huge, bushy eyebrows which moved as he spoke.

In 1933 Teller was at Göttingen, where he was quickly told that he had lost his job because of the Civil Service Laws.[27] At least he knew where he stood: some of his Göttingen colleagues had to try to second-guess what their employer would eventually decide. Quickly he joined the wave of emigration. After short spells in England, and the nuclear sanctuary of Bohr's institute in Copenhagen, he moved to George Washington University, Washington DC. For the rest of his life, Teller lived with scientists. With few other friends, science was his only contact with the rest of the world. The competitiveness which had antagonised Teller's Budapest teachers and his classmates could also irritate his physics contemporaries.

As well as working with other Hungarian exiles, notably Theodor von Karman and George Placzek, Teller also teamed up with another emigré, the flamboyant George Gamov. Born in Russia in 1904, Gamov studied in Leningrad before moving to Göttingen, then Copenhagen, where he had first met Teller, and Cambridge. After returning to Russia, he decided he had made a bad move, and after several bold but unsuccessful clandestine attempts to flee the country, left Russia for good via its front door, while officially attending a 1933 science meeting in Brussels. He made his academic home with Teller in George Washington University. There, the pair worked on nuclear theory. Gamov was trying to extend what was then known about nuclear physics to imagine what happened in the fiery nuclear furnace of the Sun's interior. This was the physics of light nuclei, mainly hydrogen, rather than ponderous uranium. This work on solar physics with Gamov sparked Teller's interest in what would become his life's work.

On arrival in the USA, Teller had been given a few hints on how to behave in his new environment. Sometimes he took note. After the new nuclear insights from Gamov and Teller, the University of Chicago sought out Teller for a job, but the authorities at George Washington University advised 'if you [just] want a genius ... get Gamov ... [but] Teller is much better. He helps everybody. He works on everybody's problem. He never gets into controversies or has trouble with anyone'.[28] After the hiatus of the war, in 1946 Teller finally went to Chicago, where he was indeed very popular.[29] At other times, Teller must have forgotten the initial advice he had been given. Apart from his spells at George Washington University and Chicago, few would have written such a recommendation. But a still well-behaved Teller was nearby when the transatlantic neutron fission news arrived with Bohr at New York City's docks. Szilard, Teller, and Wigner became known to American scientists as 'the Hungarian conspiracy'.[30] Together, they provided valuable scientific input to the new US uranium effort, but, as newcomers to the USA, and with little idea how much anything cost, they had little notion of how much cash to ask for to support their project. Their own salaries already seemed a lot of money.

Soon another influential emigrant arrived. Emilio Segrè (1905–89), born near Rome, was an Italian Jew (the name points to Spanish origins) who moved from studying engineering in Rome to physics research there with Fermi. Even before the promulgation of Mussolini's race laws, Segrè moved to Palermo in Sicily, where he had passed the selection *concorso* to inherit a professorship.[31] There he felt cut off from science. However, in the summer of 1936, he visited the University of California at Berkeley, near San Francisco, where the young US scientist Ernest Lawrence had developed his cyclotron and knew how to use it to obtain beams of subnuclear fragments much more powerful than those of the natural radioactivity used by European scientists. Before returning to Italy, Segrè arranged that material exposed to beams in Lawrence's cyclotron would be mailed to him in Palermo. This 'manna from Berkeley' suddenly brought Palermo to research prominence.[32] Segrè and his colleagues carefully extracted traces of an element that later came to be known as technetium, with 43 protons, and which filled one of last remaining gaps in the Periodic Table of the elements.

In 1938, Segrè returned to the USA, where he met Leo Szilard in New York. When quizzed about his trip, Segrè said that he was planning to return to Palermo afterwards. Szilard scoffed.[33] Mussolini would soon announce Nazi-style race laws, he informed.[34] Szilard's inbuilt predictor was working well: just a few hours later, Segrè learned of Mussolini's decision. He proceeded overland to Berkeley, aware that he would not be using the return half of his transatlantic ticket. In California, Glenn Seaborg had helped to transform Lawrence's cyclotron device into a research tool. Segrè joined him to show that bombarding uranium with energetic particles could synthesize new elements with even more protons than uranium, a man-made appendix to Nature's Periodic Table of the elements. One of these, plutonium, with two extra nuclear protons, was even more susceptible to fission than uranium 235. While Lawrence's machine could synthesize atoms of these new elements one by one, the technique would need a long time to accumulate enough plutonium to handle. Visiting his old colleague in 1940, Segrè had suggested to Fermi that the newly discovered synthetic nuclei could be bred from uranium by smothering it with neutrons, for example in the assembly that Fermi was building,[35] and then separated out by chemistry.

But it soon became clear that the Columbia University premises were simply not big enough to assemble the amount of material needed to achieve a self-sustaining chain reaction. Work switched to what had been a squash court at the University of Chicago's sports complex. In November 1942, assembly began of what Fermi had now officially called his 'pile', a roughly spherical assembly 8 m in diameter, which eventually contained some 35 tonnes of graphite, 40 tonnes of uranium oxide, and 6 tonnes of uranium metal.[36] Cadmium rods could slide in to mop up neutrons, or be rammed in if the chain reaction looked like getting out of control. It was a big project with a lot of design options, which Leo Szilard thought he knew how to solve. But he was not formally in charge of the rapidly growing scheme, and seemed to have difficulty accepting that knowledge and insight were not the same as authority and responsibility. One month later, with Fermi carefully orchestrating the absorber rods, the neutron yield was coaxed above the critical level—in fission terms, the reactor was beginning to smoulder. A celebratory bottle of Italian wine was produced, but Szilard did not drink any. The first man-made chain reaction in natural uranium was well under control, but Szliard was concerned that its soft nuclear glow could one day become an explosion. The nuclear bomb that H.G. Wells had written about in 1914 in his prophetic *The World Set Free,* based on the fictional element carolinium, was no longer science fiction.

With Fermi's prototype nuclear reactor rushing towards completion, Wigner had been put in charge of a parallel theoretical group, exploring the physics of nuclear reactors and looking ahead to future possibilities. What had been two separate themes in Wigner's career now began to come together—his initial university training as a engineer, and his new expertise in nuclear physics. With Fermi's pile having shown the way, the next logical objective would have been simply to generate heat to power turbines and produce electricity. Wigner's task was instead to design bigger nuclear reactors to breed plutonium. By the time Fermi's reactor in Chicago went critical, Wigner had drawn up detailed plans. The installations were huge. A lot of water would be needed to cool the reactors, so they were built on the banks of the Columbia river in remote Washington state.

By 1942, the Germans too had realized the fission potential of plutonium,[37] but no such grand schemes were planned to produce plutonium or other fissionable material. Instead there were other priorities—annihilating the Jews, and developing 'vengeance' missiles. Had the Germans known just one more thing, it could have turned out very differently . . .

NOTES

1. Brown, A., *The Neutron and the Bomb, A Biography of James Chadwick*, Oxford University Press, Oxford, 1997, p.27.
2. Campbell, p.386.
3. Pais IB, p.397.
4. AIP oral history, M. Goldhaber, 4632.
5. Szanton, p.78.
6. Pais IB, p.266.
7. Pais IB, p.265.
8. Szanton, p.159.
9. Rhodes MAB, p.291.
10. New York Times, 28 January 1939, reproduced in Hahn, O., *Mein Leben*, Brückmann, Munich, 1968.
11. Dahl, p.203.
12. Rhodes MAB, p.285 and Fischer, J., *George Placzek*, CERN Courier, September 2005, p.25.
13. Rhodes MAB, p.297.

14. von Weizsäcker, C., talk at CERN, 1988 '*The Meaning of Quantum Theory*'.
15. Griehl, Manfred. *Das geheime Typenhandbuch der deutschen Luftwaffe.* Wölfersheim-Berstadt, Podzun-Pallas Verlag, 2004.
16. Jungk, p.91.
17. Rhodes DS, p.27.
18. Gowing, p.50.
19. Segrè, p.113.
20. Harteck archives, Rensselaer Polytechnic, Troy, NY, cited in Dahl.
21. Szanton, p.205.
22. Rhodes MAB, p.314.
23. Weisskopf, p.188.
24. Szanton, p.199, compresses these developments into a single meeting at which Wigner was present.
25. Szanton, p.123.
26. Teller, p.48.
27. Beyerchen, p.30.
28. Letter From M. Tuve, quoted in Hargittai, p.84.
29. Jack Steinberger, private communication.
30. Rhodes MAB, p.308.
31. AIP oral history, 4876 E. Segrè.
32. Letter from Segrè to Lawrence, 19 January 1938, in the Lawrence archives at Berkeley, Cited in ref 55, Chapter 6, p.166.
33. AIP history, 4876.
34. Segre, E., *A Mind Always in Motion*: Autobiography, University of California Press, Berkeley, 1993, p.132.
35. Rhodes MAB, p.352.
36. Rhodes MAB, p.436.
37. Bernstein, *Plutonium*, p.79.

10

For in much wisdom is much grief

In any history of the Atomic Bomb[1], one document stands out because of its importance and its brevity. Before it was written, there was no clear vision: scientists were vaguely aware that a new nuclear era was about to begin but were unclear about its implications. They thought nuclear fission could only happen with many tonnes of uranium, which, as Einstein had intimated to President Roosevelt, could produce a bomb, 'carried by boat' as it 'might prove too heavy for transportation by air'. And would this fission turn out to be an earth-shaking explosion or just a slow fizzle? A few carefully composed but smudged pages banged out on a portable typewriter and headed 'strictly confidential' blew away the confusion, pointing to the definite possibility of a compact bomb and highlighting its military and political implications. Although written by scientists, it was not a scientific paper full of jargon and equations. It was meant to be understood immediately by whoever looked at it. Compared to the speculative waffle forwarded to President Roosevelt under Einstein's name, the 'Frisch–Peierls memorandum', written at the University of Birmingham in March 1940, focused attention as few other documents written by scientists had done before. In 1938, Otto Frisch and his aunt Lise Meitner had been the first to understand the fission process. In 1940, Frisch's co-author was Rudolf Peierls, another young physicist of Jewish origin who had learned his trade in Germany before leaving in haste in 1933. With Peierls, Frisch had his second uranium epiphany.

In the 1930s, Niels Bohr's institute in Copenhagen remained a vital nuclear physics clearing house, a forum where new developments were debated and fresh information exchanged. Already an architect of the quantum atom in his younger days, the mature Bohr became the father-figure of nuclear physics: it was as though everything that happened inside the nucleus had to have his authority. After 1933, his institute became a first port of call for scientific refugees, a haven of sanctuary where harassed researchers could find a secure, if temporary, base while keeping abreast of fast-moving developments, political as well as scientific.

On a smaller scale, but no less important for those who benefited from it, the warm hospitality of Rudolf Peierls and his wife Genia went on to play a similar role. Even without political upheavals, the nomadic life of aspiring young academics is often difficult, migrating from one assistantship or fellowship to the next without having time to put down roots, but at the same time having to combat challenging administrative and logistical problems: getting work permits in order to find somewhere to live, but at the same time finding somewhere to live in order to qualify for work permits. And all the time doing good research to qualify for the next job. Young scientists are like athletes, continually called upon to produce their best under difficult and unfamiliar conditions. Genia Peierls, from Leningrad, who had herself to raise a family under such trying conditions, understood. These kindnesses and hospitality fostered a huge amount of goodwill towards the Peierls. *Bird of Passage*, Rudolf Peierl's charming autobiography, describes a benevolence that extended over more than a quarter of a century. His skill in adapting to new environments, assisted by his sheer intellectual ability, provided a role-model for other migrants, some of whom lived in his house.

Peierls was born in Berlin in 1907 into a comfortable family: his father was the managing director of a prospering electrical-cable concern. His Jewish roots had been camouflaged by his having been formally baptized as a child, a move deemed expedient rather than spiritually fulfilling. Like many of his future nuclear physics colleagues, his initial ambition had been to be an engineer, but he was dissuaded because of poor eyesight and inherent clumsiness. Instead he went to study physics in Berlin, where he heard lectures by Max Planck ('the worst I have ever attended'[2]). Perhaps because of this, and by the youthful urge to leave the protective family nest, he transferred to Munich, where Arnold Sommerfeld's lectures were instead a model of clarity. Here Peierls met another future fugitive, Hans Bethe, with whom he was to have a long and productive partnership.

In 1928, Peierls moved on to Leipzig, where the young Heisenberg was establishing a dynamic new group. There Peierls worked on quantum magnetism with Heisenberg's first research student, Felix Bloch, another future refugee and a Nobel laureate. Peierls' next assignment was in Zurich, to work with Wolfgang Pauli. But he returned to Leipzig to complete his doctorate as his stay in Zurich had been too short to satisfy the university requirements. He left a strong impression on the caustic Pauli, who said of Peierls 'he talks so fast that by the time you understand, he is already saying the opposite'.

In 1930, Peierls continued his nomadic academic wandering, moving through Holland and making the statutory call on Bohr in Copenhagen before attending a conference in Odessa, where he met Eugenia ('Genia'). As a Russian in Western Europe, she had to become adept at making even more social adjustments than her husband. The pair initially settled down to married life in Zurich. While Switzerland goes out of its way to be friendly to tourists, its foreign residents can get entangled in a web of rigid formalities. Having a Soviet wife added extra spice to this administrative mix, giving the Peierls a foretaste of what was to come, both for themselves, and for the families of many of the scientists who Peierls himself would hire.

Although Peierls and Pauli got on well, Pauli always liked to have fresh faces to shout at, and in 1932 Peierls had to move on, this time armed with a one-year Rockefeller Foundation fellowship. His friend Bethe had used such a grant to cover six months in Cambridge, followed by six months in Rome, both very dynamic places for nuclear physics at the time. Peierls decided to follow suit, and was in Cambridge when the news of the Civil Service Law came. He had not been surprised, and had meanwhile turned down a job in Hamburg. He knew that his childhood baptism was not enough. In Cambridge, Peierls now began to take on his own research students, one of the first being Fred Hoyle, later to become famous as a scientist, author, and communicator.

But with the Rockefeller money soon to run out, Peierls had to move on again, this time to Manchester, where he worked for a year with Bethe, making important contributions to the understanding of nuclear physics. Peierls' bid for a permanent job at Manchester was foiled by the sudden appearance of another enforced emigrant, Michael Polanyi. Thus, in 1935 the Peierls family moved back to Cambridge, which was still jostling with non-Aryan emigrant scientists. In 1937 came an unexpected invitation to become Reader in Applied Mathematics in Birmingham, a city which, apart from the war years, was to remain the Peierls' home until 1961. Knowing that money was a concern for a family man, the offer helpfully pointed out that it would be a 'far better position than the one you have at Cambridge at present'.[3] One of the first lodgers in the Peierls' new house was

Otto Frisch, who had been invited to Birmingham in 1939 by the Australian Mark Oliphant, another Rutherford disciple who had moved away from Cambridge to a senior position elsewhere. Hiring Frisch was just one of a series of highly prescient Oliphant decisions. Frisch had packed his bags immediately, leaving the accoutrements of several years of Danish existence behind.[4] Soon war broke out, and Frisch's heavier stuff had to stay where it was. Another result of war being declared was that he and his landlord Peierls became 'enemy aliens'. Peierls' naturalization papers were slowly working their way through the British system, but his exalted academic status saved him from the cruel internment that others, less fortunate, had to endure. However there were still drawbacks: he was not allowed to run a car, and initially was not permitted to go out at night. In the anxiety of those early war years, the two small Peierls children were evacuated to foster parents in Canada.

In March 1940, Birmingham was a long way from the nuclear action in Fermi's New York laboratory. While Fermi and Szilard were building their uranium structure, not knowing how much uranium they would need to amass before anything happened, Peierls had developed a mathematical model to simulate the effect of neutrons in uranium. Feed in the appropriate reaction rates, turn the handle, and it spat out the critical amount which would be needed to make neutrons self-generating. In those early days, estimating the reaction rates of uranium was inspired guesswork. But Peierls now knew about uranium 235. If a neutron, any neutron, hit such a nucleus, most of the time something had to happen. Feeding this into his equation, he was stunned by the number which dropped out. For uranium 235 to go critical, only about a kilogram of it would be needed. And uranium is very heavy, 70 per cent heavier than lead, so a kilogram is only about a handful! He summoned Frisch, and they duly worked out what would happen. How quickly did the neutrons multiply? Would a mass that small get so hot that it would simply evaporate before the neutrons had finished doing their work, or would it hold together long enough to explode? Feverishly they calculated. They were astonished: the handful of uranium 235 would create an explosion equivalent to several thousand tonnes of Alfred Nobel's dynamite. To underline the dramatic change of scale, they called it a 'super-bomb'.[5] It would be a sphere of uranium 235 made of several parts, each too small to be critical, which would be slammed together to initiate the explosion. This would happen within a second.[6] In a single paragraph, the paper ran through the short calculation which gives the dynamite equivalent of a uranium 235 explosion. As well as direct physical damage from the blast, the memorandum pointed out the threat of lingering radiation damage.

But could it happen for real? While only a small amount of uranium 235 would be needed, this material nevertheless makes up less than one per cent of natural uranium. How could such a tiny fraction be separated out? Isotopes—nuclei differing only by their neutron content—are chemically the same, and cannot be separated in a test-tube. The only distinguishing feature is that uranium 235 nuclei are fractionally lighter. To separate them out would mean whirling uranium around in some kind of centrifuge—mechanical or electro-magnetic—or exploiting the fact that particles of different sizes diffuse through a fine mesh at different rates. The Frisch–Peierls memorandum suggested turning uranium into a gaseous compound, such as the hexafluoride, and letting it drift upwards by convection. The lighter gas rises faster, so would accumulate there.

Oliphant in Birmingham was by default the godfather to the scheme, which became entrusted to a committee obscurely named MAUD, a name taken from the cryptic 1940 telegram sent by Lise Meitner to London at Bohr's request.[7] Ironically, because of their enemy-alien status, neither Frisch nor Peierls was initially able to contribute to the work of the committee set up to investigate their suggestion. In addition, at the outbreak of war the scientific expertise of British universities had been hurriedly marshalled into research and development work, where radar was seen as a major priority, while the prospect of a uranium bomb, remote and obscure, remained sidelined in university departments.[8]

One who helped was Franz Simon, born into a wealthy Berlin Jewish family in 1893 and who served in the artillery during the First World War, where he was one of the first poison- gas casualties on the German side. Twice wounded, he was decorated with the Iron Cross First Class. His scientific career focused on cryogenics, working initially in Berlin with Walther Nernst. (The ageing Nernst, who had two Jewish sons-in-law, had no quarrel with Jews, but did with Nazis.) In spite of being exempt from the 1933 Civil Service Law because of his war record, Simon left his homeland in 1933. Like Schrödinger and the London brothers, he moved to Oxford at the invitation of Lindemann. Before leaving Breslau, where he worked at the *Technische Hochschule*, Simon was ordered to surrender his passport, which would have prevented him leaving. Instead, he flung down his Iron Cross and demanded whether he was supposed to surrender that as well. His furniture removal was overseen by a negligent official who was kept well supplied with beer, and who at one stage left behind the briefcase containing his orders, which said that Simon's 'goods have to be closely inspected since he is a dangerous and subversive character who could spread anti-Nazi propaganda'.[9]

In Oxford, Simon continued his research in cryogenics, but with the outbreak of war and the increasing focus on nuclear physics, switched direction. Simon still had strong feelings about Nazis. To Born, he wrote that he was resolved to 'use my whole force in the struggle for this country'.[10] To underline his assumed new allegiance, he became Francis instead of Franz. Peierls knew Simon of old, and felt that he was the best man to lead an effort to separate uranium 235 by diffusion. For this, a crucial quest was to find the right material to filter uranium hexafluoride gas. If the microscopic holes were too small, the process would be too slow. If they were too large, it would no longer be a filter. Simon found a way by hammering a domestic tea-strainer. Oxford attracted several other exiles. Nicholas Kurti was another talented Hungarian Jew whose science career in Berlin had been interrupted in 1933. There, he had been working under Simon, and both of them were happy to be able to continue this tradition in Oxford. With Simon and Kurti, Peierls had the makings of a team—refugees all—to investigate how to separate uranium 235.

PO Box 1663

In the USA, Szilard had continually pushed and manoeuvred. In spite of the clarity of their insight, Frisch and Peierls in the UK had no political plan. Theirs was a voice from afar. However influential British scientists visited the USA, and their American counterparts increasingly frequently went to London. In July 1941, after more than a year of pushing, the implications of their work were grasped across the Atlantic.[11] From this point, US fission research and development work took off. The story of the Atomic Bomb, like the life of Einstein, has been told many times, and another detailed account is superfluous. It was an unprecedented flood tide of scientific achievement as well as a turning point in world affairs. For the first time, scientific capability was amassed to such an extent that, when pushed into motion, like an avalanche, it became irresistible. The quantum exiles who had made it happen became caught up in this momentum, and many were surprised at where it took them. Only after the avalanche had run its course were they able to see where they were. Only a few had remained on the sidelines: Einstein, although a figurehead and a prominent spokesman, had little to do with the war effort;[12] Pauli, his colleague at Princeton, likewise; Schrödinger remained in Dublin in neutral Ireland; Max Born, in Edinburgh, had strong pacifist feelings.

The project needed so much support and logistics that the US Army was brought in, with Brigadier-General Leslie Groves in command. Groves immediately threw a cloak of secrecy over the operation, now called the 'Manhattan Project'. Its expansion had been so rapid that it looked in danger of becoming fragmented. Fermi and Szilard were building a prototype nuclear reactor in Chicago; scientists at Berkeley were looking at plutonium, the chemistry group delegated to advise on the production of plutonium was led by James Franck at Chicago. The man who had helped develop poison gas for the Germans in one war was now helping to develop a mightier weapon to deter them in another. Various schemes were being proposed to separate uranium 235 from 238. To oversee and guide all the different arms of this huge scheme needed a new scientific focus, a big one. A new laboratory town, dedicated to nuclear research, fission development, and bomb manufacture, would be built at an almost undeveloped site at Los Alamos in New Mexico, code-named 'Site Y', or for communications purposes PO Box 1663, Santa Fe. Having chosen the site, someone had to be chosen to run it.

At the beginning of the twentieth century, subatomic physics had been a European affair. Discoveries were made in Paris and Berlin, and later Rome, with Rutherford's laboratory at Cambridge for a long time enjoying some kind of monopoly. The understanding of the quantum processes inside the atom was dominated by German-speaking scientists, with the notable exception of Bohr in nearby Copenhagen.[13] Very little happened across the Atlantic. Aspiring American quantum scientists had to come to Europe to serve their apprenticeship. One was J. Robert Oppenheimer, the talented son of a sophisticated and successful Jewish family. To study physics, he first went to Göttingen, working under Born and meeting the new generation of quantum pioneers, and later to Zurich to work with Pauli. Coping with the unfamiliar European lifestyle and the challenges of quantum physics led to deep psychological problems: Pauli complained that Oppenheimer seemed to spend more time with the psychoanalyst than doing science. Returning to the USA before the discovery of fission, Oppenheimer brought the European mindset to the USA and implanted it in new schools of US quantum expertise in California. This had borne fruit even before the arrival of the exiles in the mid-1930s. Oppenheimer's first generation of American research students[14] went on to become very influential.

Oppenheimer, a cultured visionary, was appointed scientific director at Los Alamos. His next task was to recruit nuclear experts for the new laboratory-town. An élite had already arrived in the quantum exodus from Europe. Some of them Oppenheimer had met while he had worked in Göttingen and Zurich. Others were already involved in the project. In the summer of 1942, Oppenheimer brought six senior scientists together at Berkeley: three of them—Hans Bethe, Edward Teller, and Felix Bloch—were fugitives from Europe. The other three were Americans: John Van Vleck, who went on to receive the 1977 Nobel Physics Prize for his work on electrons in magnetism, Robert Serber, and Emil Konopinski.

Hans Bethe (1906–2005) —'The Prophet of Energy', according to his biographer Jeremy Bernstein—was born in Strasbourg, or rather Strassburg as it was called at the time, as it had been in Germany since 1870. (It would remain so until 1918, when the name reverted to the French form.) His father was a physiologist with a liberal political outlook. Bethe followed his father's political leanings, but from the outset his interests and abilities were clearly more mathematical. His father was initially bemused by his son's ambition, but later became intrigued by it. His mother was Jewish, and felt that her son should embark on a career that would be profitable. However Frau Bethe's vision of an engineer son did not make much headway. After studying physics at Frankfurt, Bethe moved on to research in Munich under the sagacious guidance of Sommerfeld. Then, with a Rockefeller award, he moved on, first to Cambridge, where he learned English, and then to Rome with Fermi, where he did not learn Italian. Returning to Germany in 1932, he was given a temporary appointment in Tübingen, near Stuttgart.

In a town he did not know, Bethe felt alone, especially as it was a Nazi hot-bed—'one of the most Nazi-infested towns of Southern Germany'.[15] In April 1933 he learned of the Civil Service Law. Although he realized that it concerned him, having a Jewish mother, the first direct news came in a letter from one of his students, who had read in his local newspaper that Bethe had been sacked from Tübingen. Bethe immediately wrote to his superior there, a man with whom he had had good relations, for clarification. A stiff reply confirmed what the local newspaper had said, followed by a letter from the local Ministry of Education saying that Bethe's salary would end soon.

On 11 April 1933, he wrote to Sommerfeld, 'As so often, I have to ask your advice: you probably do not know that my mother is a Jewess, so therefore according to the new laws I am 'not of Aryan descent' and so not worthy of being an official of the *Deutsches Reich*. ... Where my 'birth mistake' (*Geburtsfehler*) is known and from where, I have no idea'.[16] Returning to Munich, Bethe was sheltered for a few months by Sommerfeld, but there were many Nazis in Munich, and Bethe learned that he had to keep his opinions to himself. He relaxed when he returned to England, first to Manchester, then in Bristol, where his colleague was Walter Heitler. When he left Germany, Bethe was able to take out almost a thousand marks. (His future wife, Rose Ewald,[17] who left three years later, was only able to take ten marks).[18] Bethe enjoyed England and would like to have stayed, but only temporary jobs were available, and Bristol did not want 'to stand in his way if there was a more remunerative offer from overseas'.[19] However, his short stay nevertheless seemed to catalyse his interest in nuclear physics: 'I imagine that after a few more years, I would have been captivated [anyway] by nuclear physics, but it came earlier ... because England was full of nuclear physicists when I came there in 1933'.[20] In 1935 he was invited to Cornell University, Ithaca, New York. There, he created a dynamic new physics group and discovered a lot about nuclear reactions, including the detailed mechanisms which fuel the Sun's nuclear furnace. For this he was awarded the Nobel Prize for Physics in 1967.

With the German army overrunning much of Europe in 1940, Bethe felt obliged to do something. The United States was not yet in the war, but after talking with Théodor von Karman he decided to team with Edward Teller to understand how shock waves could make high-speed missiles penetrate armour plate, their private contribution to the war effort. As a nuclear physicist, Bethe was aware of the new bomb speculation, but was sceptical of its feasibility.[21] In spite of this, Oppenheimer had been impressed by Bethe's ability to solve problems, whether understanding the Sun's nuclear fuel or designing projectiles to drill through armour plate, and had plans for him.

For his think-tank, Oppenheimer had also selected Felix Bloch (1905–83), born in Zurich to Jewish parents who had emigrated from Austria. Bloch was also pointed towards studying engineering at Zurich. But there, like von Neumann, he was led into physics by Peter Debye, Hermann Weyl, and Erwin Schrödinger. When von Neumann departed for Berlin in 1927, Bloch went to work with Heisenberg in Leipzig. After periods in Copenhagen with Bohr and in Rome with Fermi, he returned to a teaching post at Leipzig in 1932. Bloch had lived with German anti-Semitism, but had initially not taken it seriously. That all changed in 1933, and Bloch simply left Leipzig,[22] even more informally than Schrödinger had left Berlin. Schrödinger at least had told them he would no longer be teaching. In spite of his unannounced departure, Bloch was implored to return, being promised guards in the lecture room to prevent any trouble. That did not impress Bloch, who felt that the guards themselves would have been the first cause of any unpleasantness. As a Swiss, Bloch liked to climb mountains. Once he slipped and fell, hanging by a rope held by a Swiss physicist colleague, Egon Bretscher. Bloch survived to receive a job offer from Stanford University, a place he had not heard of, but assumed was in the United States because the salary was in dollars.

Oppenheimer offered Bethe the key job of Head of the Los Alamos Theory Division, one of the most dazzling assemblies of brain power ever put together (it included the young Richard Feynman, who quickly bonded with Bethe). It was not a university department where researchers were free to do whatever caught their fancy. Instead, this unsurpassed collective of intellectual vigour had to be marshalled and focused on specific problems. One of their allotted tasks was to freeze-frame what happened in explosions, microsecond by microsecond. The Theory Division headship was the job that Teller had assumed he would get. As though being overlooked for the plum job was not enough, Teller now nursed another grievance. Physicists had now seen that uranium fission was not the only route to nuclear energy. There was another, potentially even more powerful than anything uranium 235 or plutonium fission could produce. Instead of heavy nuclei being induced to split, light nuclei had to be fused together to make heavier ones. It was the process which made the Sun shine and emit life-giving warmth. But with the fission route fast-tracked at Los Alamos, the fusion option was sidelined. Teller was doubly miffed.

Los Alamos was huge, but nominally top secret. Researchers were only supposed to discuss what really concerned them, and not to pry into the work of others. This was anathema to many university intellectuals, who were used to free discussions with whoever they wanted. And all were supposed to be US or UK citizens. Becoming a US citizen takes time, and many European emigrant recruits had not yet accomplished these administrative moves before the US formally declared war on Germany, a few days after the Pearl Harbour attack in December 1941. As had happened already to their colleagues in Britain, they technically became 'enemy aliens', prohibited from owning a short-wave radio and excluded from war work.[23] To move to Los Alamos, each 'alien' first had to have two sponsors to vouch for them, and were told to keep their new assignation secret.

Victor Weisskopf (1908–2002) was another product of a highly cultured Jewish family in Vienna. At the age of just 20, he appeared at Göttingen, there to do research with Born and Franck, then with Wigner in Berlin. Climbing up the academic ladder, he moved to Leipzig to work with Heisenberg, Bloch, and Placzek. In 1931 he moved back to Berlin, where he saw at first hand what the Nazis were capable of. With a peripatetic Rockefeller award, Weisskopf travelled briefly to Kharkov, and then to Copenhagen and Cambridge, before becoming Pauli's assistant in Zurich, where he developed a curious reputation for blinding insight coupled with an uncanny ability to make arithmetical mistakes. His students invented a new number, the 'Weisskopfian', which in rough calculations could take the value 1, π, or ½, either positive or negative.

Weisskopf's penchant for Russian visits had been noticed. While he was working with Pauli in Zurich, he was called to the office of the Institute's President, an unusual occurrence for a mere research assistant.[24] There, Weisskopf was asked whether he had ever had any connection with the Communist party, an allegation which he could honestly deny. A few weeks later, he was summoned to the police department which dealt with foreigners, and was shown into an office where on the desk was a large folder marked *Kommunistiche Umtriebe* (Communist Activities). Inside was a copy of every letter Weisskopf had received in Switzerland. He was astonished—in an era before copying machines, every letter he had received had been opened and manually duplicated. Weisskopf tried to convince them that he was politically clean. The police told him that normally he would be immediately expelled from the country, but the Institute had pleaded that he was indispensable to Pauli. However, after his contract at Zurich officially expired, Weisskopf was supposed to leave Switzerland 'for ever'.

Moving again to Copenhagen, in 1937 his interest shifted to the detailed study of nuclear processes with Bohr.[25] After being offered a job in the Soviet Union and turning it down, in 1937 he moved instead to Rochester, New York. Weisskopf relates that European immigrant scientists were being exploited: as a married man, his initial salary of $200 per month was breadline. The same happened to Segrè at Berkeley. But they were happy to have their jobs, a long way from the problems in Europe. Still going through the preliminaries of getting US citizenship, in 1943 Weisskopf heard a shout in the corridor at Rochester University 'Professor Weisskopf, telephone from Santa Fe'.[26] By then, many people had learned what a phone call from Santa Fe meant. So much for secrecy. When he arrived at Los Alamos, he was sternly told that his mission was 'to produce a practical military weapon in the form of a bomb in which the energy is released by a fast neutron chain reaction'.[27] The scientists soon found their own ways of paying lip service to strict rules, while continuing the discussion and collaboration so vital to their work. Some thrived at Los Alamos, others found the atmosphere of secrecy oppressive. Bloch did not like having his mail opened and being under surveillance, and left in November 1943, after just a few months. Oppenheimer was furious. Bloch went to work instead on wartime radar development.

Frisch and Peierls, whose 1940 memorandum had catalysed the project into action, were still in Britain. With Birmingham mainly geared to Oliphant's radar work, to continue his fission studies Frisch moved north to Liverpool, where neutron pioneer James Chadwick had set up an influential group. The port city had come under continual heavy air bombardment, and Frisch had to move house and his place of work several times. As an 'enemy alien' he was initially under curfew and not even allowed to ride a bicycle. However Chadwick was able to get special dispensation from the police for his protegé.

As the US nuclear effort gathered momentum, Chadwick had become a key figure. As well as coordinating the British effort, he was the only Briton with a complete knowledge of what was then going on across the Atlantic. In 1943, Chadwick asked Frisch if he would like to join the effort at Los Alamos. Frisch jumped at the chance. However, this meant becoming a British citizen. Within a week, formalities were accomplished, including an oath of allegiance to the King and formal exemption from military service. Clutching his new UK passport decorated with a US visa, Frisch boarded a ship bound for Newport News, Virginia. On board was the advance party of British brainpower for the atomic bomb project, also including Peierls, George Placzek, Felix Bloch's rescuer Egon Bretscher, and a research colleague of Peierls called Klaus Fuchs. At Los Alamos, Peierls was reunited with Bethe and with many other figures that he had met on his travels in the 1930s. When Chadwick moved from Los Alamos to Washington, Peierls became the head of the UK contingent at Los Alamos. At Christmas 1943, listeners to Los Alamos local radio heard 'Otto' (for security reasons no family names were used) playing the piano. At Los Alamos, Frisch's talent for laboratory work led him to become the leader of a group designing critical assemblies for 'hot' uranium, exploring just how much radioactive material could be assembled safely before its neutrons got out of control, in which he and others had accidental encounters with large doses of radioactivity. Frisch attributed some subsequent eyesight problems to these imprudent wartime irradiations.

Another Los Alamos recruit from Britain was Josef Rotblat, born into a comfortable Warsaw Jewish family in 1908. Unlike the other exiles, who had all lived and worked in Germany, Rotblat's background was essentially Eastern European. In Warsaw, Rotblat's father had owned a flourishing transport concern, but when all horses were requisitioned in the First World War, the proud family was ruined. His parents had marked out their son as a future rabbi,[28] but Josef was more attracted to mathematics, for which Poland had rediscovered a tradition. For the now-destitute family, formal schooling was a luxury, so the enterprising Rotblat instead became an electrician. By studying in the evenings he eventually clambered to a university degree, and by 1937 had become Assistant Director of the Atomic Physics Institute of Warsaw's Free University. When he heard the news of the discovery of fission, Rotblat investigated the production of neutrons from uranium, convincing himself that a chain reaction and all its consequences would be possible.[29] He jumped at a chance to move to Liverpool to work with

Chadwick, but his abysmal annual salary of a mere £120 could not support two, and his wife (he married in 1937) remained in Poland. In Liverpool, Rotblat, whose knowledge of English had been acquired from classical literature, had enormous problems adapting to spoken English, especially the local Liverpool variety. In August 1939, Rotblat was promoted, doubling his salary, and returned to Poland to fetch his patient wife. However she could not travel because of appendicitis, and he returned to Liverpool alone. Within a few days, Poland was invaded by Germany and his wife was unable to leave the country.

At Liverpool, Rotblat soon met Frisch. When asked by Rotblat to play the piano for some Polish soldiers stationed in Liverpool, Frisch was horrified to find an old upright piano with many keys not working. Undeterred, he pounded out Chopin's *Grande Polonaise* to thunderous applause.[30] As a key member of the UK neutron team alongside Chadwick, in 1943 Rotblat was ordered to proceed to Los Alamos with Frisch and Peierls. Rotblat, proud of being Polish, refused to take British nationality, and was initially left behind. Later he was granted exemption as a special case. Perhaps because of this, his relationship to the bomb programme differed slightly from that of his colleagues.

To strengthen the Los Alamos contingent, Emilio Segrè arrived from Berkeley to look at the exact amounts of uranium 235 and plutonium needed for bomb construction. Improved knowledge of uranium 235 properties could make the bomb smaller, and an easier load for a bomber.[31] By 1943, the total amount of plutonium available was just 1.5 milligrams. Kilograms of it were needed for a bomb, and to shape one, Franck's team in Chicago had to find out the exact properties of the new metal—its melting point, hardness, density, etc.[32] Segrè also showed that a plutonium bomb needed a different trigger mechanism. The gun and bullet technique foreseen for uranium 235 would just fizzle for plutonium. Instead, a hollow sphere of the synthetic metal would have to be imploded. After these vital contributions to the wartime bomb programme, Segrè returned to Berkeley and went on to share the 1959 Nobel Physics Prize for his role in discovering nuclear antimatter.

During this time, Niels Bohr, the nuclear Pope, was still in distant Copenhagen. There he was cut off, except for delegations from Germany led by Werner Heisenberg. In 1943, Bohr's position became very precarious when the Germans decided to reinforce their grip on Denmark and send in troops. With forewarning, most of Denmark's 5000 Jews, including Bohr, were quickly transferred by boat to Sweden. There a British bomber was sent to pick him up. In London, he brought news of Lise Meitner and others who had found refuge in Sweden. Under the assumed name of Nicholas Baker, Bohr eventually arrived in Los Alamos, where his sudden appearance galvanized its scientist inhabitants. The most recent émigré from Europe underlined the contempt for Hitler and renewed the belief of the scientific community in their objective. Although the scientists no longer needed Bohr's scientific help, his very presence inspired Los Alamos. Einstein did not go there.

As well as motivating the scientists, Bohr also badgered the politicians. Unconcerned by the technical details of bomb manufacture, he saw the post-war implications of the new nuclear world. He had been in contact with Soviet scientists, and realized that after the war, a new nuclear arms race would begin, this time without Hitler. In 1944, Bohr shuttled between Roosevelt and Churchill, sowing seeds of post-war implications. However, both leaders were preoccupied by more immediate concerns. In spite of his status, Bohr, mumbling and inarticulate, was not the best spokesman. Even for those who knew him, Bohr could be difficult to understand. After listening to Bohr's muttering for about half an hour, Churchill asked him to stop.[33]

As the bomb work progressed, Szilard became increasingly uncomfortable in a large team with a strict hierarchy run under strict military rules. Workers were not supposed to discuss technical matters which did not directly concern them. This did not suit the mercurial Hungarian, convinced that he had the solution to every problem. Szilard became a conspicuous misfit. The increasingly exasperated General Groves alleged that Szilard was 'the kind of man that any employer would have fired as a troublemaker'.[34] Szilard obstinately refused to keep his mouth shut, floating above the teeming hive of the project, a vagabond troubleshooter. Groves, piqued by these antics and less sensitive to the scientific aspects of the work, wanted Szilard removed and even interned as potentially dangerous. Still refusing to keep quiet about matters officially secret, Szilard took his case to Washington, where he was shadowed by security agents. Monitoring Szilard absorbed much valuable effort at a time when it should have been focused elsewhere. They had got the wrong man: Szilard's antics were unwittingly a smokescreen for the real espionage elsewhere at Los Alamos.

Inside Germany

As the advancing Allied armies overran nuclear installations in southwest Germany in April 1945, it soon became clear that the Nazis had not even been in the contest to build a bomb. Any race had instead been a walk-over. With the end of the war in Europe now imminent, the question facing the US government now was what to do with a bomb that was almost ready. The accumulated momentum of such a gargantuan project so near to completion was difficult, if not impossible, to stop. The most obvious target was another enemy. If Japan were threatened with such a weapon, maybe the Land of the Rising Sun would surrender without having to be invaded.

The prospect of dropping the bomb reactivated Szilard, who resolutely continued to pull strings. Just as he had engineered a letter to Roosevelt warning of the dangers of not starting a nuclear effort, so, five years later he organized another missive, this time advising Roosevelt to stop the effort before it was too late. But it lay unread when the President inconveniently died of a cerebral haemorrhage on 12 April 1945. Undeterred, Szilard scrabbled for strings to manipulate the new President, Harry Truman, while influential hawks pushed for the bomb to be used against Japan as soon as possible. Szilard helped orchestrate a committee, chaired by the prominent James Franck, pointing out the future dangers of a nuclear arms race.[35] Instead of running headlong towards nuclear deployment, the report urged that nations should strive instead to control the development of such weapons on the basis of mutual trust. Few people could yet comprehend all the talk of twenty thousand tons of TNT, so the report recommended a demonstration explosion to show the new danger facing an unsuspecting planet. Franck personally took the report to Washington. But with the war in the Pacific coming to a belligerent crescendo, these arguments were not heard. Instead, another panel, including Oppenheimer, advised that the bomb should be used for real.[36] Fermi was also on that panel, the only non-native American, but the quantum exiles who had convinced the United States to build the Atomic Bomb and helped them to build it were about to leave the stage.

As the Allied armies advanced across Germany, primitive nuclear installations were not the only things they found. There had been whispers throughout the war, but few had heeded. After witnessing what was going on, in July 1943 Jan Karski of the Polish resistance had gone to

Washington to spread the news that the Nazis had begun mass executions of Jews. He had a fruitless meeting with Roosevelt. Supreme Court Justice Felix Frankfurter, himself born in Vienna into a Jewish family, which had left for the USA at the turn of the century, could not believe what he heard. As the eminent lawyer later explained, 'I did not say that this young man was lying: I said I was unable to believe him. There is a difference'.[37] Some two years later, as Soviet armies advanced across Poland, and British and US troops entered Germany, disbelief in Karski's revelations turned to stunned realization.

The 1933 Civil Service Law had been the thin end of a wedge which Nazi Germany slowly hammered into its defenceless Jewish population. A slow crescendo of persecution reached an initial frenzy in the *Kristallnacht* riots of 1938. Their declared aim was to force Jews in Germany to emigrate,[38] and it was successful: by 1939, the Jewish population of Germany had shrunk by two-thirds, from half a million to 175 thousand. But in 1939, German troops overran Poland, where there were more than three million Jews. Half a century before, in Tsarist Russia, millions of Jews had been harassed into emigration by pogroms, but this had taken twenty years. Hitler could not wait that long. With *blitzkrieg*, the status of Jews under Nazi rule changed.

Hitler had warned what was coming in January 1939, in his traditional Reichstag speech marking the anniversary of the Nazi accession to power. If other countries criticized Germany for expelling Jews, he mocked, 'Why were the others not grateful for the gift that Germany was giving the world. Why do they not take in these magnificent people?' His sarcasm suddenly switched to outright threat: 'I have often been a prophet, and I have mostly been laughed at.... It was mostly the Jews who laughed.... Today I want to be a prophet again. If international Jewish financiers... should succeed in plunging the nations into world war... the result will be the annihilation of the Jewish race in Europe'.[39] As they drove deep into Poland, the Nazis rounded up Polish Jews and herded them into ghettos, conveniently situated near major railheads. In Warsaw, the largest, with a population of 450,000, the conditions were so bad that over 40,000 died in 1941 alone.[40] Had such attrition continued, all would have disappeared in ten years. But the pace of Hitler's fury was still accelerating. He invaded an unsuspecting Soviet Union, home to another several million Jews. What should the Nazis do with them all? This was the 'Jewish problem'.

For the Germans, nuclear fission, stripped of talent and lacking visible advocates, had been pushed into the background. Development work was moving ahead for what would become the *Vergeltungswaffen* (revenge weapons)—the V1 cruise missile and the V2 ballistic missile—but this was not yet seen as a priority item. Meanwhile there were more important objectives. On 20 January 1942, a high-level meeting, 'followed by luncheon', at the headquarters of the Criminal Police Commission in the Berlin suburb of Wannsee examined the 'final solution' to the 'Jewish problem'. Construction had already begun in Poland of new mass-extermination centres, but the carefully stage-managed meeting ensured that the enormous logistics of moving and killing so many Jews became a national priority, even for a nation totally committed to war. The meeting originally had been scheduled for December 1941, but was postponed after the German army's failure to capture Moscow, and with the Japanese attack on Pearl Harbour, US President Roosevelt had become the leader of a nation at war. Continually pressured by nuclear lobbyists, on 19 January 1942, just one day before the Wannsee meeting, he became finally convinced that the big effort to develop a fission bomb should be given special priority.[41]

The two sides had committed themselves to their respective vast schemes, each sucking off huge quantities of resources, and absorbing attention. While one was initially seen as a deterrent to destruction, the other was a rampant aspiration to slaughter. The respective logistics of what became the Atomic Bomb and the Holocaust were both challenging, but very different. They went their own ways, and on different timescales. There was no sense of selection, only of direction. But making such decisions was channelled by the level of resources available. Whatever else, total commitment can never exceed the sum of what is available.[42]

By the end of 1942, when the Los Alamos site became the centre for atomic bomb development, some four million Jews had already been exterminated. In 1943, as new priorities for revenge weapons and other new projects emerged, concentration camp internees were also seen as a convenient source of slave labour, leased by the SS to government industries for a few marks per day.[43] By the end of 1943, when the first traces of plutonium were being extracted from the new Oak Ridge reactor,[44] another 1.5 million Jews had perished. In the summer of 1944, as plutonium began arriving at Los Alamos, and the advancing Soviet army overran the Majdenek extermination camp, further west, another half a million had been massacred. In 1945, as Soviet troops advanced across Poland and the Allied armies entered the German heartland, the world was stunned to discover that Karski's message had been believable. But the Atomic Bomb was not yet totally credible, even by those who were trying to make it happen.

On 16 July 1945, one of the scientists assembled at the desolate Jornada del Muerto in the New Mexico desert was Otto Frisch. Stationed some 35 kilometres from the first Ground Zero, 'fearing to be dazzled and burned'[45] at 05:30 he stood with his back to the firing site. 'Suddenly, and without any sound, the hills were bathed in a brilliant light'. For hundreds of miles around, unsuspecting witnesses, less well prepared, thought the Sun rose twice on that day. The first sunrise had not been a soft glow in the East, but a blinding flash, accompanied by thunder which never seemed to stop. An army munitions depot had blown up, local inhabitants were told later.

The nuclear material detonated at Jornada del Muerto had not been the uranium 235 suggested by Frisch and Peierls. Instead, it was the new synthetic nuclear explosive, plutonium, manufactured in Wigner's giant reactors. Human reactions are conditioned by experience, but nobody watching that event had any idea what to expect. All were amazed by the unfolding vision. Oppenheimer, the intellectual, quoted Sanskrit and remembered the legend of Prometheus. Bethe's scientific brain was more prosaic, freeze-framing the spectacle in physics terms: 'The isothermal sphere at the centre of the expanding fireball continues opaque and invisible, but gives up its energy to the air ... by radiation transport'.[46] Fermi remembered to drop pieces of paper confetti as the blast wave moved past, and from watching where they fell was able to gauge the energy release. Such were the reactions to a milestone in world history: scientists had reproduced conditions that existed only in the genesis of stars. Apart from giant meteors crashing in from outer space, never before had such a mighty thing happened on Earth. And it was man-made.

There was a brief collective euphoria that such an unprecedented collective effort had reached its objective. But any initial triumph was soon damped. While the human race struggled to understand that atomic bombs had been dropped on Hiroshima and Nagasaki, those in charge contemplated the future of warfare. The explosions may have hastened the end of the war, but they also tilted the world into a new era. Weisskopf later quoted Ecclesiastes: 'For in much wisdom is much grief, and he who increases knowledge increases sorrow'.[47] The initial jubilation of the Los Alamos scientists sublimated into guilt. They did not want Hiroshima and Nagasaki to happen again, and sought to influence the development of nuclear policy, both in the USA and through international channels. Caught up in the scramble to achieve their objective, few of them had yet had time for such thoughts. Bohr had been a lone voice that nobody could hear. In an attempt to defuse a muddled international situation, in August 1945 the US government published *The General Account of the Methods of Using Atomic Energy for Military Purposes*, written by Henry Smyth of Princeton. It was basically a handbook for the Atomic Bomb, omitting some vital technical details.

The total cost of the wartime Atomic Bomb project has been estimated at $1.9 billion, the majority of which went into the construction and operation of the huge plants to provide fissile material.[48] This figure is comparable with the wartime expenditure in the United States on conventional bombs, or on small arms, and a fraction of that spent on armoured vehicles or on aircraft. Thus the Bomb was a small, but visible, item on the military expenditure sheet. To put this in context in another way, in 1952, the new Federal Republic of Germany (West Germany) agreed to pay reparations to the equally new state of Israel equivalent to 3 billion deutschemarks,[49] about half the cost of the Atomic Bomb. Thus a stripped-down wartime German nuclear project might have been feasible, had it been prioritized. But Hitler had other objectives seared into his mind. On 29 April 1945, from his bunker in Berlin, his final testament ranted that the war had been 'provoked exclusively by those international statesmen who were either of Jewish origin or worked for Jewish interests' Those responsible had paid for their guilt 'by more humane means than war'.[50]

Spies, real and suspected

The biggest unknown in the postwar equation was the role of the Soviet Union. Some had prophesied political climate change. Winston Churchill was no longer British Premier, but still a visionary. In February 1946, he spoke of an 'Iron Curtain' which would cut Europe in two. Bohr had warned that Soviet scientists were more than curious about the Atomic Bomb. His warning was borne out in the autumn of 1945 when Igor Gouzenko, a code clerk at the Soviet embassy in Ottawa,

Canada, walked in from the street and asked for political asylum. At first, few wanted to see him, but once contact was made, he eventually revealed how the supposedly secret wartime nuclear effort in the USA had been riddled with leaks. The Soviets already knew far more than what was in the Smyth Report.

Security for the Manhattan project had been problematic, with hard-pressed agents, unused to dealing with Europeans, continually having to chase after a scientific sprint. General Groves had been worried about the sudden influx of scientists speaking foreign languages. But the quantum immigrants had their influential patrons—except Szilard, who did not care. In all this, with an Eastern European background and his refusal to take British nationality, Rotblat was conspicuous. As if to confirm these suspicions, in Santa Fe, Rotblat had also been noticed seeing a woman, Elspeth Grant, an old friend from the UK. He had also been taking flying lessons.[51]

At Los Alamos, Rotblat had initially believed that he was working against the clock to beat the Germans. But when he saw the immense effort underway in the USA, and even before German installations had been overrun by the advancing Allies, he realized that a wartime Germany under continual air bombardment would be incapable of matching this. There was no reason for him to be at Los Alamos. Behind the science, he also discerned sinister political goals: the Soviets, allies in the war against Nazi Germany, had already been identified in the USA as a long-term threat. Shocked, disillusioned, and worried about his family, Rotblat boldly left Los Alamos before the first bomb was tested.

For the security people, there was only one explanation. En route to New York for the boat back to England, all Rotblat's personal possessions mysteriously vanished. The allegation was that, after returning to Britain he would fly to Poland and pass nuclear intelligence to the eagerly waiting Soviets.[52] Back at his old post in Liverpool, Rotblat indeed made contact with Poland, but for another reason. He learned that his wife and her mother had perished in Majdenek concentration camp.[53] He never remarried.

After Elspeth Grant came to visit Rotblat in 1950, she was interrogated on her return to the USA. Now hypersensitized by what had happened elsewhere, the US authorities were convinced that Rotblat was a spy. For years, he was denied a US visa. For much of the rest of his life, the highly principled Rotblat campaigned vigorously against the spread of nuclear weapons, and went on to share the 1995 Nobel Peace Prize for his work with the Pugwash movement for nuclear disarmament. He died in 2005, aged 96, soon after Hans Bethe, aged 98, the last survivors of the quantum exodus. But for the US security service, Rotblat had been a red herring: the biggest fish of all slipped through all the security nets.

Academia is a strange business. In university departments or scientific research centres, talented people from very different backgrounds are thrust together. Some naturally get along, others do not, but the intensity of their mutual interest can usually overcome social and cultural barriers of all kinds. Academics get used to immigrants or visitors who do not know the rules of their new environment, and make allowances for them. In climbing the scientific ladder, young researchers have to clamber through a series of short-term assignments. Candidates have to be selected quickly on the basis of ability, gauged by reputation and recommendation. If promising young researchers are seen as having something to offer, they are hired. If they are good, they climb the ladder. If the selection process is wrong, their brief probationary contract is not renewed, and they slip off their precarious perch. For the wartime nuclear physics boom, there was a vast range of research to be covered and a shortage of brainpower to do it. The authorities in Britain and the USA had initially been reluctant to entrust sensitive work to 'enemy aliens', but in the rush to get the fission programme underway, security requirements were short-circuited.

Most of the quantum refugees left Germany because of their Jewish origins, or, like Schrödinger, because of personal antipathy toward the Nazis. But Nazi intolerance extended to other areas. Initially, communists were more undesirable than Jews. Their fate, harsh imprisonment, was worse than losing one's job. The graceless Klaus Fuchs, born in 1911, had studied physics at Kiel and Leipzig. His mother committed suicide via the memorable route of swallowing acid.[54] The emotionally scarred Fuchs later joined the German communist party.

Committed to his politics, Fuchs left for Britain and managed to resume physics studies, first in Bristol, and from 1937, in Edinburgh under Born, who vouched that Fuchs was 'among the two or three most gifted theoretical physicists of the young generation'.[55] At the outbreak of war, Fuchs became an enemy alien, and,without a prestigious university position, was interned in Canada. His contingent of internees crossed the Atlantic in two ships, and the one which carried all the paperwork was torpedoed. Fuchs, on the second vessel, thus arrived as an unknown quantity with nothing to refute allegations of being a Nazi agent. This did nothing to increase any goodwill he had towards Britain. However Born intervened on his behalf, and Fuchs was able to return to science in the UK, where he obtained British nationality in 1942. Born, reluctant to join the British fission bomb effort, had earmarked his pupil Fuchs instead,[56] who was sent to help Peierls with fission development work.

Fuchs became another of the academic nomads who found shelter in the Peierls household. For a nuclear job, he needed official clearance from the British authorities, who were now ready to accept Peierls' recommendations. Fuchs was fast-tracked for British citizenship. When he informed Tess Simpson at what had been the Academic Assistance Council, she replied that she was 'full of delight'. But the unfazed researcher, realizing the implications of fission work, had also made contact with Soviet envoys in London. In 1943, he crossed the Atlantic again, to become a key member of Peierls' team, initially working on uranium 235 separation on the East Coast, but when Peierls was summoned to Los Alamos, Fuchs went too. There he became well integrated into the life of the laboratory town, getting drunk at parties and always ready to lend his car to Richard Feynman at weekends. Scientists noticed his dedication, even when others were working seventeen-hour days. On arrival in the USA, it had not been difficult for him to re-establish contact with Soviet agents. After he moved to Los Alamos in August 1944, and was at the centre of bomb development, it was more difficult for him to maintain contact with Soviet agents in New York. They were worried when he suddenly disappeared from view. In February 1945, he visited his sister in Boston and re-established contact. He had a lot of information to pass on.

After the war, Fuchs continued to have access to all papers at Los Alamos, including ongoing plans for new weapons. In 1946, he was rushed by air to the new British Atomic Energy Research Establishment at Harwell, near Oxford. There he became head of its theoretical physics group, with his long-time colleague Otto Frisch as head of nuclear physics. Harwell's aim was to get an independent British bomb programme going as quickly as possible. Travelling from Harwell to London was easier than going from Los Alamos to New York, and he continued his double game. In 1950, it was the British security service which finally caught up with Fuchs. Peierls and Frisch were astonished when their protegé was arrested on charges of 'high treason'—passing key nuclear information to the Soviets. Calling themselves 'perturbed scientists',[57] other emigrants who knew Fuchs and had become naturalized Britons—Max Born, Francis Simon, Rudolf Peierls, and Joseph Rotblat —offered to surrender their passports. As the person who had recruited him and been his boss at Los Alamos, Peierls felt particularly concerned. Researchers count on their colleagues' integrity. He had no idea what Fuchs had been doing, Only later did he discover that the dates of Fuchs' contacts with Soviet agents coincided with those of frequently declared 'sickness'.

After a trial in London lasting less than two hours, Fuchs freely admitted to being guilty of all charges. He was sentenced to fourteen years' imprisonment, and stripped of his British nationality. He felt little, if any, remorse, but admitted his abuse of friendship. He had never been 'recruited' by the Soviets and his motives were naively altruistic. For him, all sides had to share such important knowledge, and he felt justified in doing all that he did, for whatever country, and was shocked to be condemned as a criminal. Released after nine years, he was taken by police car to London's Heathrow airport to board a flight to East Germany, moving into an important nuclear research post. After receiving national honours, he died in 1988.

A very different story was that of Friedrich 'Fritz' Houtermans, who became the victim of his own conflicting allegiances, and who, unlike Szilard, contrived to be in the wrong place at the wrong time.[58] He was born in 1903 in Danzig (then part of Imperial Germany) to a Dutch father and an Austrian Jewish mother. After their divorce, Houtermans lived with his mother in Vienna, and grew up to become a motivated communist. His physics research at Göttingen (where he was a contemporary of Oppenheimer) in the mid-1920s had been noticed, and he moved to Berlin to work with Gustav Hertz. In 1930, during a trip to the Soviet Union, he married the physicist Charlotte Riefenstahl. The pair had first met in Göttingen, where she had also been courted by Oppenheimer. Wolfgang Pauli was a witness at the wedding ceremony. When the Nazis came to power in 1933, Houtermans—doubly targeted as a communist and as a *Mischling*—initially fled to Copenhagen. His wife knew many of the physicists who had managed to move to England, and Houtermans was able to move to a relatively well-paid industrial post in EMI's television development group in London. His wife had a post in Cambridge. But, keen to return to academia and still enamoured with communism, Houtermans shifted to Kharkov in the Soviet Union, where one of his colleagues was George Placzek. The scientists' imprudent criticism of the Soviet system was soon noticed, but Placzek was able to get out before it was too late. However Houtermans fell foul of Stalin's Great Purge; he was imprisoned for more than two years on a charge of being a German spy, and learned that making a false confession under harrowing interrogation immediately improved his lot.[59] While rewarding to those who both gave it and extracted it, such false information was ultimately less helpful to higher authorities, who knew full well what went on. Houtermans was eventually invited to say who had forced him into making his 'confessions'.[60] While this was going on, his family, now in London, continually pounded on the doors of those with political influence. Blackett in Britain, the Joliot–Curies in France, and Bohr in Denmark all tried to apply high-level diplomatic pressure for his release.

With the signing of the short-lived Molotov–Ribbentrop Soviet–German alliance, Houtermans was released in 1940 and taken to the German border, where, with his far from satisfactory political background, he was immediately imprisoned by the Gestapo. Released through the strenuous efforts of Max von Laue, he became a remote part of the nascent German nuclear effort. When the Germany army invaded the Soviet Union in 1941, Houtermans, with his knowledge of the Ukraine, had gone with them, ostensibly to earmark Soviet science worth importing into Germany. This was interpreted by many of Houtermans' German colleagues as collaboration with the Nazis. However while there, Houtermans also shielded and fed former Soviet colleagues who would otherwise have been rounded up and taken to concentration camps.[61] Rejoining the disorganized periphery of the German nuclear physics effort, he saw the potential of the route to fission through plutonium, and engineered messages to the USA hinting at what was afoot.[62] His 1942 telegram 'Hurry up, we are on the track', sent via Switzerland, reached Wigner in Chicago[63]. This message has often been the subject of speculation. Was Houtermans actually warning the Allies, or was he trying to fool them? It could have been a carefully engineered follow-up to Heisenberg's visit to Bohr in Copenhagen in 1941.[64] Whatever the explanation, Houtermans was continually effervescent. He was the saviour of Richard Gans. This and other altruistic efforts appeared to escape the attention of the Nazis. However unable to get enough cigarettes, the resourceful

Houtermans managed to procure 50 kilograms of waste tobacco from cigarette factories, possibly for the purpose of carrying out experiments on the physiological effects of nicotine for the military, possibly as a source of heavy water. This finally got him fired.[65]

While all this was going on, his wife made her way to the USA, where she became more visible to the Allies than her husband deep inside Germany, and was the subject of high-level concern on both sides of the Atlantic. After the war, Houtermans returned to academia, first at the intellectually impoverished Göttingen, then in Switzerland, where he turned his attention to geophysics. One of his contributions was a calculation of the age of the Earth from a detailed analysis of meteoric remains. During their long enforced separation, the Houtermans had become divorced under wartime law. When contact was re-established, they remarried in 1953, Pauli again being a witness. However their second marriage did not last. With a colleague and under assumed names, Houtermans wrote of his harrowing experiences in Soviet prisons.[66] As well as his exploits, he is remembered through the Houtermans medal awarded annually to outstanding young scientists by the European Association of Geochemistry. There is also a lunar crater named after him.

NOTES

1. The chapter heading refers to Ecclesiastes, 1, 18.
2. Peierls, p.19.
3. SPSL, Box 335.
4. Rhodes MAB, p.319.
5. Not to be confused with Teller's later thermonuclear 'Super' bomb.
6. For the full text, see e.g. Gowing, Appendix 1.
7. See Chapter 8.The telegram ended 'Please inform Maud Ray Kent'. Dahl, P., *Heavy Water and the Wartime Race for Nuclear Energy*, Institute of Physics Publishing, Bristol, 1999, p.118.
8. Gowing, p.37.
9. Arms, N., in SPSL, Box 339.
10. Rhodes MAB, p.339.
11. Rhodes MAB, p.368.
12. other than occasionally being a consultant for the Naval Bureau of Ordnance, see Chapter 7.
13. and Paul Dirac in Cambridge.

14. together with those of John Van Vleck.
15. Hans Bethe, AIP oral history, 4503.
16. Eckert, M. and Märker, K. eds, *Arnold Sommerfeld, Wissenschaftlicher Briefwechsel,* Band II (1918–1951), Deutsches Museum, Berlin, 2004, p.380.
17. the daughter of Paul Ewald, see Chapter 4.
18. Bernstein, *Bethe*, p.40.
19. SPSL, Box 324.
20. Eckert, M. and Märker, K. eds, (see note 16), p.355.
21. Bernstein, *Bethe*, cited in Rhodes MAB, p.415.
22. Felix Bloch, AIP oral history, 4510.
23. Weisskopf, p.116.
24. Weisskopf, p.120.
25. J. D. Jackson, *Research Highlights*, CERN Courier December 2002, p.8.
26. Weisskopf, p.122.
27. Rhodes MAB, p.460.
28. Biographical Memoirs of Members and Fellows of the Royal Society Vol 53 (2007), Hinde, R. and Finney, J., p.312.
29. Underwood, p.90.
30. Frisch, p.138.
31. Rhodes MAB, p.541.
32. Los Alamos Science, Vol 23 (1995), p.162.
33. Jungk, p.161.
34. Rhodes MAB, p.502.
35. Jungk, p.169.
36. Jungk, p.170.
37. Wood, E.T. and Jankowski, S., *Karski, How One Man Tried to Stop the Holocaust*, Wiley, New York, 1994.
38. Friedländer, p.287.
39. Arad, Y., *et al.* (eds), *Documents on the Holocaust*, Yad Vashem, Jerusalem, 1962. Quoted in Friedländer, p.310.
40. Engel, p.52.
41. Rhodes MAB, p.388.
42. In quantum mechanics, this seemingly innocuous limitation, called by physicists 'unitarity', can be highly consequential and produce unexpected effects.
43. Cornwell, p.342.
44. Rhodes MAB, p.548.
45. Gowing, Appendix 5.
46. Rhodes MAB, p.672.
47. Jacob, M., in CERN Courier, Weisskopf commemorative issue, December 2002, p.19.

48. Hewlett, R.G. and Anderson, O.E. Jr., *The New World: A History of the United States Atomic Energy Commission, Vol 1, 1939/1946* (Oak Ridge, Tennessee: U.S. AEC Technical Information Center, 1972), pp. 723–4.
49. American Journal of International Law, 48(4), October 1954.
50. Gilbert, M., *Second World War*, Wiedenfeld and Nicholson, London, 1988, p.677.
51. Hinde and Finney, see note 28.
52. Underwood, p.22.
53. Another who just managed to escape from Warsaw in 1939 was Roman Smoluchowski, a versatile physicist equally at home when studying atoms, stars, or the Moon's surface. Born in 1910 in Zakopane, then in Austria-Hungary but now in Poland, he studied at Warsaw and Groningen, and worked briefly with Einstein in Princeton in the 1930s before returning to Warsaw. In the USA, he worked at the Carnegie Institute, General Electric, Princeton, and Texas.
54. Rhodes DS, p.54.
55. SPSL, Box 328.
56. Jungk, p.105.
57. Sunday Express, London, 29 October 1950.
58. Landrock, K., Friedrich Georg Houtermans (1903–1966) – Ein bedeutender Physiker des 20. Jahrhunderts, *Naturwissenschaftliche Rundschau,* Vol 56, No 4, 187–9 (2003).
59. Amaldi, p.625.
60. Amaldi (note 59), p.629.
61. SPSL, Box 330.
62. Bernstein, *Plutonium*, p.95.
63. Szanton, p.241.
64. This was the subject of the highly successful play *Copenhagen*, by Michael Frayn. The 1942 Houtermans telegram could also be a compelling subject.
65. Freund, P., *A Passion for Discovery*, World Scientific, Singapore, 2007, p.125.
66. Beck, F. and Godin, W., *Russian Purge and the Extraction of Confession*, Hurst and Blackett, London, 1951.

11

Science and anxiety

When J. Robert Oppenheimer brought his hand-picked sages together in the summer of 1942, fission bombs were the main item on the agenda. But the assembled erudition soon started discussing something very different from a bomb made of uranium or plutonium. Nature's 92-fold abundance of chemical elements also hid another nuclear effect. The neutron had been predicted to explain the heaviness of nuclear matter. However nuclear masses are not simply the sum of those of their component protons and neutrons. Even in large, unstable nuclei like uranium, the combined effort of its nuclear particles gripping together is enough to pare off mass according to Einstein's equation, $E = mc^2$. When some of these loosely bound neutrons and protons slip out of each other's grasp in nuclear fission, the remaining ones grip together even more tightly. Lise Meitner had quickly seen the implications: the consequent nuclei are so happy to see their unwelcome tenants go that they release the fireworks of fission.

At the other end of the Periodic Table, the mass patterns work in the reverse direction: light nuclei become progressively more tightly locked together, and energy is released whenever light nuclei are synthesized. For example helium, with two protons and two neutrons, is marginally lighter than these constituents on their own. If these protons and neutrons could be coaxed to form helium, then part of their mass would volatilize into energy. After Houtermans had first toyed with the idea, in 1939 Bethe finally wrote down the solution to a long-standing puzzle: what seemingly inexhaustible supply of fuel has kept the Sun shining for billions of years? Positively charged protons, pushed apart by their electrical repulsion, find it difficult to get close enough for their nuclear grip to find a handhold. However at temperatures of millions of degrees, such as those inside the Sun, the nuclei hurtle around fast enough to overcome this electrical hostility. As its seething protons collide, the Sun's hydrogen fuses into helium 'ash', releasing sunshine; in physics-speak it undergoes 'thermonuclear fusion'. The process is self-generating as long as the supply of hydrogen lasts. Ironically, such fusion had been seen in the laboratory six years before the discovery of fission: in 1932, Oliphant at Cambridge first saw what he called *Transmutation effects with heavy hydrogen.*[1] But that is typical of science: discoveries do not always happen in a logical order.

In 1941 Fermi had idly suggested to Teller that perhaps thermonuclear fusion could be made to happen elsewhere than just in the Sun and remote stars.[2] An Atomic Bomb, already the greatest release of terrestrial energy ever, could spark an even bigger explosion. The nuclear mass sums said that any such bomb would make a fission bomb look puny. Having launched the monstrous idea, Fermi seemed to forget about it. However Teller did not. The prospect caught fire in his mind, and he could think of little else. At Oppenheimer's 1942 brainstorming session, Teller locked onto the idea. He had already relegated the Atomic Bomb to an engineering problem. As Bethe moved into his new job as Head of the Los Alamos Theory Division, Teller was ordered to stop thinking about a fusion bomb. But he was unhappy at being passed over for the post now filled by Bethe, and could not stop thinking about what he now called a 'Super' bomb. He no longer pulled his weight as a fission team player, and annoyed other Los Alamos scientists by playing his grand piano at night. Some of the fission-related work he should have been doing was given instead to Klaus Fuchs, who didn't seem to mind.

Among all the talent assembled at Los Alamos, arguably the biggest single increase in cerebral power came with the arrival of John von Neumann. Great minds like Einstein ranged widely, but nevertheless stayed put long enough to paint scientific pictures for posterity. On the other hand, von Neumann darted from one problem to another, like a bee in a garden, searching for the richest source of nourishment, pollinating flowers as it goes. His contributions resemble more a full-length stream-of-consciousness film than the more static images left by Einstein and Gödel. But while Einstein was inspiring and Gödel baffling, von Neumann in the flesh was dazzling, his brain constantly exploding like an unpredictable firework display. In contrast to the remoteness and impenetrability of other scientists, von Neumann was an extrovert who enlivened any occasion. Wearing a trademark smart three-piece suit and carefully folded handkerchief in his breast pocket (even while riding a horse), von Neumann influenced everything he did and impressed everybody he met. He could handle English, French, German,

Hungarian, Yiddish, Greek, and Latin and had a seemingly inexhaustible supply of jokes and erudition in most of these languages.

The childhood of von Neumann János, as he was first called, traced a very different path through the turbulence of 1918 Hungary than that of the young Edward Teller. The von Neumann background was already a major influence: the son of a successful banker, he was part of a comfortable extended family which lived together in a large house. With German and French-speaking supervision, the various children—von Neumann's cousins—were naturally multilingual. But the young von Neumann quickly outshone everyone. At the age of five, he impressed his pushy parents by revealing that he knew the Budapest telephone directory by heart. As if that were not enough, he could reverse-search it in his head: given any telephone number, he could immediately say to whom it belonged.

Inheriting an ability which had been seen before in the family, von Neumann could also do mental arithmetic with numbers too long for others even to remember. His memory was palatial: his brain was imprinted with every book of interest he had ever read, and he could recite them from memory, if necessary translating them into other languages as he went. On the shelves of the family library in Budapest had been the 45 volumes of world history edited by the German academic Wilhelm Oncken. By the age of ten, Von Neumann had assimilated the contents, and drew on this knowledge for the rest of his life, constantly astounding his audience. But his memory had inbuilt levels of priority: while he could remember dates in Byzantine history, he could not always remember what he had for lunch, or the details of his diary and assignments.[3] For those less gifted, it is easy to confuse intellect and memory. Producing facts off the cuff at parties or in TV quiz shows, or doing mental arithmetic is impressive, but is not in itself intellect. This is more difficult to put on show: Einstein could not do party tricks, but von Neumann could not play the violin. His friends were also embarrassed by his habit of staring at every woman in sight.

As well as learning under Laszlo Ratz, with Eugene Wigner, von Neumann had also been taught by Gabor Szego, who went on to lecture in mathematics at Berlin and Königsberg. (As Szego was Jewish, that changed in 1933, when he moved to the USA, becoming prominent in the mathematics department at Stanford University.) The mathematical career suggested by Ratz and Szego was resisted by von Neumann's father, who preferred his son to turn towards something more mundane. Through the influence of the pioneer aeronautical engineer Theodor von Karman, the young von Neumann was pointed instead towards a career in chemical engineering. He complied, and signed up for study, initially in Berlin, then Zurich, but adding mathematics in Budapest.[4] For the latter, he showed up only for the examinations, but that was enough. At the age of 26, von Neumann moved to a teaching post in Berlin, soon also earning a Rockefeller award for post-doctoral work with Hilbert at Göttingen, where he met Oppenheimer. Von Neumann's teachers soon sensed that they were dealing with someone special. Hermann Weyl, when asked a difficult question during a Zurich seminar, spotted von Neumann in the audience and asked him to supply the answer instead.[5]

Von Neumann's brain saw things differently. While others were taught to understand the world around them from the bottom up, as an assembly of components, each with intensity, substance, and energy, von Neumann instead saw a top-down global picture full of structure, organization, and controlling forces. While others found the new quantum mechanics incomprehensible, von Neumann cast it into a rigid new mathematical formalism. Redirecting his mathematical powers, von Neumann also pioneered game theory as the calculus of strategy, and developed the ideas of programming for the new digital computers then beginning to appear. As his reputation spread, for several years in the early 1930s, von Neumann shuttled between posts at Princeton and in Europe, but this stopped in 1933, and he had to sever his ties to Germany. Before taking up his first invitation to lecture in the USA, he married a Catholic girl, and became Catholic himself, making an effort to moderate his strong language.

When Oppenheimer assembled his brainstorming group in the summer of 1942, von Neumann was not included, nor did he become a permanent Los Alamos inhabitant. People knew that he could solve difficult problems quickly, and having him in full-time residence was an unnecessary luxury. He was one of the few exempted from the security rule that all involved in the project had to live on the site. Los Alamos would simply call on him whenever necessary, and they were not disappointed. He was a genius at handling the complex mathematics of explosions. After intense intellectual sessions, Los Alamos scientists would often relax by playing poker for small stakes. Von Neumann could apply his own strategy theory at any time, but rarely took the game seriously. Once he lost five dollars, and the winner of the game stuck the bill inside a copy of von Neumann's classic book (with Oskar Morgenstern *et al.*) *The Theory of Games and Economic Behavior.*[6] The book's highly erudite chapter 'Poker and Bluffing' is incomprehensible to most poker players.

The emergence of strategy as a discipline was helped by the US Air Force's creation of the RAND (Research and Development) Corporation to follow on from wartime operational research. RAND became the prototype think-tank for post-war technological issues. Von Neumann was a consultant, brought in from time to time to advise on problems that baffled other people. By this time, his belief in his own ability could verge on the arrogant. He was once asked to help in the design of a new computer to handle a specific problem. After being briefed on the problem, von Neumann concluded, 'Gentlemen, you do not need a new computer: I have solved the problem'.[7] On another occasion, von Neumann discussed with a colleague the mathematical approach to a problem in operational research. During the night, the colleague mulled over the advice, and gradually realized its worth. Excited by the insight it brought, he wanted to congratulate von Neumann, but knowing that von Neumann was a late riser, waited until 10 the next morning before making the phone call. 'You wake me early in the morning to say that I am right!', complained the genius, 'of course I am right! Next time only wake me if I am wrong'.[8]

Von Neumann's contributions in the area of digital computing are among the best remembered. He was an adviser to several major computer manufacturers when their first computers appeared: many of the fundamental design decisions were his: stored programs running on digital machines based on binary arithmetic. This involvement had begun with the quest for calculation power, with improvements in the design of the pioneer wartime ENIAC (Electronic Numeric Integrator and Calculator) machine to adapt it for post-war work at Los Alamos. This was followed by inconclusive efforts to build his own computer at Princeton. He also advised on the design of new machines for work on nuclear strategy and defence issues.[9] In 1954, he had to rein in his wide-ranging activities when he was appointed a member of the US Atomic Energy Commission. In these later years, von Neumann's increasing involvement in private enterprise and in the exercise of power stifled the scientific integrity that had characterized his earlier work.

In 1954, Fermi, just 53, died of cancer. After Los Alamos, he had moved to Chicago, where he motivated a new generation of researchers. The disease that killed him could have been due to a lifetime of imprudent exposure to radiation. The following year, von Neumann was also diagnosed with cancer. As the disease spread, it caused him agonizing pain. One of his final projects had been to use the human brain as a model for computer architecture, but the drugs administered to dull the pain also clouded his mind. Instead of reeling off literature, he mumbled gibberish. His hospital room was guarded in case he blurted out nuclear secrets. On his deathbed in 1957, at the age of 53, a priest formally converted him to Catholicism, in case there had been any misunderstanding.

Von Neumann's parents had deemed that their son's mathematical talents were incompatible with a career. Only through his own insistence did his ability blossom. However, a neighbouring country was enjoying an intellectual revival in the early twentieth century. Long dominated by its neighbours, Poland re-emerged as a separate nation after the First World War. Mathematics became an important focus of the Polish revival, continuing a tradition set by Nicholas Copernicus four hundred years before. As well as the abstract work of Stefan Banach and Alfred Tarski (born Alfred Teitelbaum, a Jew who converted to Catholicism), in 1934 a group of Polish mathematicians, seconded to a special unit, broke the codes for German message encryption machines. This gave the Allies a valuable head start when the Second World War came. Also anticipating the threat of a German invasion was the Mandelbrot family, who moved from Warsaw to Paris in 1936. The young Benoit Mandelbrot, then aged 11, went on to discover fractals, bringing a new visibility and beauty to mathematics.

Into this Polish mathematical revival came Stanislaw Ulam, born in Lviv in 1909, the son of a lawyer. In those days, Lviv, in Galicia, was part of the Austro-Hungarian Empire, only becoming Polish in the post-1918 rearrangement of frontiers. In his autobiography, *Adventures of a Mathematician,*[10] Ulam curiously makes no reference to any Jewish background until he relates meeting other Jewish scientists in later life, where they are able to exchange erudite Yiddish aphorisms. When Lviv was threatened by Russian forces in 1914, the Ulams migrated to Vienna, but returned in 1918. Stanislaw's interest in science had been aroused by the 1919 solar eclipse observations, which confirmed Einstein's new picture of relativity. At Lviv, Ulam sat for hours in cafés discussing mathematical problems. By 1933, he had a doctorate in mathematics. Although Poland was not directly affected by Hitler's accession in Germany, the implications were clear, and Ulam's family urged him to continue his studies further afield, bankrolling him through visits to Zurich, Paris, and Cambridge. On return to Poland, a letter awaited him with the offer of a job at Princeton. It had been arranged by von Neumann, whom Ulam had not yet met, but each knew of the other's mathematical reputation.

Ulam's modesty is everywhere in his autobiography, but he could not hide his ability. Through his reported dealings with his colleagues, we learn that he had a phenomenal memory, and had absorbed Greek and Latin as well as a clutch of modern European tongues, together with the family Yiddish. In the late 1930s, Ulam shuttled between posts in the USA and his native Poland, managing a timely exit just before war broke out in Europe. Noticing that US scientists seemed to be mysteriously disappearing from the academic scene, Ulam asked von Neumann about the possibility of war work. The result was a letter from Bethe offering Ulam a job in a mysterious establishment of which he knew nothing, other than that it was in New Mexico. Going to the university library to consult books about the area, he discovered they had all been borrowed earlier by other scientists who had vanished. So much for secrecy.

On arrival at Los Alamos in late 1943, the first scientist Ulam met was von Neumann, who explained what was going on. When Ulam arrived, von Neumann had been talking with another scientist who Ulam did not yet know. This was Teller, to whom Ulam was initially assigned. After the earlier introduction to fission research from von Neumann, Ulam was further briefed by Teller, who could talk about nothing other than his thermonuclear ideas. While other people at Los Alamos were busy doing other things, Ulam was soon deep in calculations more to do with nuclear fusion than with fission.

Fusion confusion

In 1939, nuclear physicists in the Soviet Union had been as aware as anyone else that fission was important. As with the Germans, their interest had been piqued when they noticed that scientific papers about fission in standard transatlantic journals like *Physical Review* stopped appearing. After the German invasion, Soviet fission research moved east of the Ural mountains. But this clumsy effort, starved of resources, suddenly changed gear in August 1945. After the US atomic bombs were dropped on Japan, there was another bombshell when Stalin heard the news. The Soviet bomb effort became top national priority, lubricated by information pumped in from Fuchs.

The evocative term 'Cold War' had first been used in an obscure 1945 article on the Atomic Bomb by the visionary writer George Orwell. Its frostiness increased after the Soviet blockade of West Berlin in 1948, becoming cryogenic when the first Soviet Atomic Bomb was exploded at Semipalatinsk in August 1949. There had been no foreign observers, and the event was detected in the West only through routine US long-range weather reconnaissance over the Pacific. A sensitive airborne radioactivity detector picked up a signal several times larger than 'ALERT'.

The USA had resigned itself to the Soviets eventually getting their own fission bomb, but were nevertheless shocked by the rapidity of this response. They did not yet know how much it had been helped by Fuchs and others. It had also been assisted by German nuclear scientists, among them Gustav Hertz and Manfred von Ardenne,[11] who moved temporarily to the Soviet Union after 1945.[12]

While Stalin's catch-up effort gathered momentum, the US nuclear community, its initial mission seen as accomplished, had largely dispersed.[13] Bohr returned to Copenhagen, and the British contingent, including Fuchs, went back from whence they came. Oppenheimer became Director of Princeton's Institute for Advanced Study, and chairman of the General Advisory Committee of the new US Atomic Energy Commission, which later took over nuclear matters from the military. The US weapons effort slowed to a crawl, the gentle accumulation of a stockpile. Little attention was given to how to deliver these weapons.

With Los Alamos lacking direction, and with no clear route to the development of his thermonuclear bomb, Teller preferred to leave in 1946 and return to academia, this time in Chicago.[14] In April of that year he had chaired a private cabal, attended by von Neumann and Ulam. Its conclusion was that Teller's 'Super-bomb' was feasible, provided it was given the necessary resources. Teller would have liked to recruit Bethe, but this man was now a 'concerned scientist' who wanted to cut his ties with weapons development.[15] In addition, his group at Cornell included Richard Feynman, another Los Alamos alumnus, who was now charting a new way of understanding quantum electrodynamics, a science that was to be as significant for the second half of the twentieth century as Einstein's work had been for the first. Feynman and Bethe revered each other.

As the Cold War intensified, Los Alamos alumni were fazed by aggressive Soviet moves in Czechoslovakia and Hungary, countries they remembered. It reminded them of the Hitlerian hostility which had presaged one war. Now there was a new perceived threat. Von Neumann and Teller had themselves seen the mayhem that a communist takeover had produced in Hungary, earlier that century. Teller, whose parents still lived in Budapest,[16] had been coaxed back from Chicago to Los Alamos by the time the psychological impact of the first Soviet nuclear explosion was felt. In January 1950, the USA committed itself to a crash nuclear development programme for the second time in a decade. This time the engineering problem was to develop a thermonuclear bomb, and Teller was convinced that he knew how to go about it.

He saw two routes. One, called 'Alarm Clock' (because it would wake up the world), used layers of light nuclear material sandwiched inside a large imploding fission bomb (this was similar to the solution that the Soviets would employ). The other route was Teller's favourite: his long-advocated 'Super', a huge tank of heavy hydrogen to be fusion-detonated. While acknowledging that smaller Alarm Clocks would work,[17] Teller pushed his Super alternative, because it would give a bigger bang. After a spell at the University of Southern California, and a major illness, Ulam returned to Los Alamos to join Teller's effort. Numerical calculations, already difficult for fission, now became both more complex and more important. With digital computers still rudimentary, Ulam stumbled on a new approach. Rather than embarking on calculations that were impossible anyway, his idea was to simulate the outcome by injecting random numbers into the calculation and watching how the 'results' mapped out. This became known as the 'Monte Carlo' method, a mathematical casino which produces results. (To see how it works, imagine some irregular shapes torn from a large sheet of paper. The objective is to compare their areas, which are difficult to calculate. A Monte Carlo approach is to scatter a handful of centimetre-square confetti over the shapes, then count how many squares randomly fall inside each one.) The technique became much more powerful when digital computers became available. At first, these had to be wired up by hand for each computational task, but von Neumann soon invented the idea of stored programming, writing down logic to execute each required task. It was the first computer software.

Ulam's Monte Carlo calculations and von Neumann's computer techniques showed that Teller's dream of a Super-bomb would fizzle rather than explode. This was underlined by another round of calculations by Ulam and Fermi.[18] A tank of thermonuclear fuel would be difficult, if not impossible, to ignite with a fission bomb: energy would escape faster than it could reproduce itself in the chain reaction. The nuclear conditions inside the bomb assembly had instead to be prepared more carefully. It would explode only if the thermonuclear fuel itself were compressed, as well as heated, by a fission explosion. Teller, crestfallen, found it difficult to accept that the idea he had clung to for almost a decade was deficient, especially when the objections depended on the roll of mathematical dice. Only grudgingly did he accept that the compression route was required,[19] followed by some fast footwork to claim credit for the new development.[20]

Development now looked to be set on a firm path, but there was still a problem. Just as had been the case in the fission project, Teller, despite his new authority, was difficult to fit into a large team effort. When he was passed over as head of the thermonuclear programme, and offered instead an apologetic assistant directorship, Teller walked out in a final thunderclap of frustration. With this, the main thrust of US nuclear weapons development finally cut loose from the 1930s' generation of quantum exiles. With this umbilical cord severed, and with the growing suspicion that its hospitality and indulgence had also been a Trojan horse for enemy agents, the USA began to slam shut the doors of tolerance which had been opened a decade earlier.

'Foreign-born agitator'

Einstein played no direct part in the effort which produced the Atomic Bomb, but, as he said himself, 'The first Atomic Bomb destroyed more than the city of Hiroshima. It also exploded our inherited, outdated political ideas'.[21] The enormous visibility of the Bomb, together with Einstein's fame, coupled with some vague awareness that he had something to do with the science that made it work, led many to consider him as its spiritual father. He refuted these allegations categorically: 'My part in it was quite indirect [and] I did not foresee it'. When he had been told by Szilard that uranium could support a chain reaction and release energy, Einstein replied '*Daran habe ich nicht gedacht*' ('I never thought of that').[22] Einstein's misgivings about quantum mechanics meant that he had been left behind in the advance of nuclear physics in the 1930s. Later, out of touch with even more new developments, he became a scientific anachronism, but he still commanded immense respect. Everywhere. His letters to Presidents left their mark: his appearances among scientists at seminars or talks were greeted by reverential silence, and all would stand.[23]

His relentlessly probing mind tried to perceive what lay behind the new scientific picture. For Einstein, there had to be something deeper, something which ensured predictable cause-and-effect at any level. One of his sayings, often repeated,[24] was 'God does not play dice'. Once he amplified this to 'It seems hard to look in God's cards. But I cannot for a moment believe that He plays dice and makes use of "telepathic" means, as the current quantum theory alleges He does'. Einstein described himself as a deeply religious man. However, his personal view of religion held that science is the thin outer veneer of a masterplan incomprehensible to mere human minds. But whatever this intangible God, Einstein could not conceive of it having any power to reward or punish people, or to guarantee life after death, or many other of the conventional trappings of religious dogma. He certainly did not believe in a supernatural God who permeates the trivia of everyday life. This, he said, was the aspiration of 'feeble souls'.[25] Einstein used the concept of a supreme deity freely: *Subtle is the Lord*—the title of Abraham Pais' definitive scientific biography of Einstein,[26] comes from a quote by the scientist himself—*Raffiniert ist der Herrgott aber boshaft ist er nicht* ('subtle is the Lord, but He is not malicious'). He produced inspirational epithets—'Science without religion is lame, religion without science is blind'.[27] But such overarching ideals were for him incompatible with 'the idea of a being who interferes in the course of events'. Such pronouncements incensed those with a more traditional view of religion, rekindling the fires of 1933: 'You ridicule the idea of a personal God. In the past ten years nothing has been so calculated to make people think that Hitler had some reason to expel the Jews from Germany'.[28] Others invited him to 'go back to Germany where you come from'.[29] In religion as well as science, the incisiveness of Einstein's thought made him controversial.

In 1945, the bombshells of the nuclear age jolted him out of his reverie. When he had moved to the USA in 1933, he had been warned to keep his head down and his mouth shut. For more than a decade, he duly complied. Confidential letters to President Roosevelt had not broken this resolve. Now, as mankind grappled with the implications of the most unimaginable development that it had ever produced, he was called upon to be a voice for the conscience of science and scientists. Einstein emerged from his long silence and gave many press interviews and made invited speeches. Their subject was no longer the space-time structure of the Universe. As scientists left their wartime roles and struggled to resolve their guilt feelings, in August 1946, he was appointed chairman of the new Emergency Committee of Atomic Scientists which aspired to give these concerned academics a voice in political circles.

In his own right, Einstein also became a strong advocate of 'world government'. He was referring to the spirit behind the newly created United Nations, but he was speaking at a time and in a country where such words could imply very different objectives and invoke disproportionate responses. His once muted voice had became shrill: 'The foreign policy of the US since the termination of hostilities has reminded me ... of the attitude of Germany under Kaiser Wilhelm II.' Not content with one indiscreet comparison, he soon drew another: 'Herein lies a certain resemblance to Marxism'.[30] Einstein became increasingly tactless, poking his nose into matters that were not his business. In 1951, Julius and Ethel Rosenberg were sentenced to death for their role in passing atomic secrets to the Soviets. Einstein, ever the pacifist, pleaded for their sentence to be commuted. In 1953, with academics and intellectuals increasingly under fire from reactionary politicians, he advocated that they should refuse to appear if called upon to testify at congressional inquisitions, choosing his words badly—'I can only see the revolutionary way of non-cooperation'.[31] In all the spheres of human activity, the passions aroused by politics and religion are the most volatile. Now Einstein freely dabbled in both, and reactionary thunderbolts were hurled back. 'One of the greatest fakers the world ever knew is Albert Einstein, who should have been deported for his communist activities years ago. The bunk he is now spreading is simply carrying out the communist line'.[32] Uncomprehending but vociferous US legislators also condemned him: 'This foreign-born agitator would have us plunge into another European war ... to further the spread of Communism throughout the world'.[33]

Ever since his arrival in the USA, Einstein had been watched by the Federal Bureau of Investigation, who eventually amassed a 2000-page file on him.[34] After 1945, when Einstein's voice became more noticeable, the FBI had a lot more to work on. Even the dignified mainstream press berated him as a meddler. On 13 June 1953 a *Washington Post* editorial admonished 'Einstein ... has put himself in the extremist category by his irresponsible suggestion. He has proved once again that genius in science is no guarantee of sagacity in political affairs'. Some thought otherwise. On 9 November 1952, Israel's founding President, Chaim Weizmann, died. The nation's Prime Minister, David Ben Gurion, cabled his ambassador in Washington to ascertain whether Einstein, the most prominent Jew in the world, would be willing to accept the nation's Presidency. Over the telephone, an agitated Einstein explained his reluctance. The new nation had become embroiled in controversies and in confrontations with its Arab neighbours. These were matters he did not feel qualified to handle. He also knew he was ill: an operation in 1948 to remove an abdominal growth had uncovered an aneurysm. In 1955 it killed him.

The nuclear explosions over Hiroshima and Nagasaki had indeed thrust civilization into a new era, one for which nobody was prepared. In 1944, Bohr had tried to warn Roosevelt and Churchill, but people either could not understand him, or did not want to. The ambivalence of the wartime bomb—the triumph of victory coupled with a chilling foreboding of cataclysmic dangers to come—was mirrored in the changing relations between the Western Allies and the Soviets, who transformed almost overnight from comrades in arms to bitter enemies. Hanging over this dilemma was the question of the future of the whole nuclear enterprise. How should the United States manage and develop its new expertise? Belligerent voices clamoured to retain the existing military infrastructure, with all its apparel of secrecy. Others, including many of the scientists who had been involved, saw the goal shifting to the peaceful exploitation of nuclear energy, under civilian, rather than military, control. But these scientists were crossing from territory they knew well into a vast political game for which they did not know the rules. Nevertheless, they were still perceived as heroes who had helped to win the war, and influential people were ready to listen to them. In Washington, the nuclear scientist lobby led to the formation of a Senate Special Committee on Atomic Energy. Its scientific advisor was one Edward Condon.

Vigorous work by this committee prepared the ground for the Atomic Energy Act of 1946 (often called the McMahon Act after its sponsor), which brought a new civil authority, the US Atomic Energy Commission, to oversee national nuclear developments. In a complete turnaround after the strict secrecy surrounding the wartime work, the plan now was for nuclear matters to be shifted into the public domain, where they could be openly monitored. When it was first opened for debate, the Act proposed 'the free dissemination of scientific information and for maximum liberality ... for related technical information'. The Act's stated objectives were 'that the development and utilization of atomic energy shall, as far as practicable, be directed toward improving the public welfare, increasing the standard of living, strengthening free competition in private enterprise, and promoting world peace'. But this was all subject to 'the paramount objective of assuring the common defense and security'. On the international front, the 'Baruch Plan' (after US financier and Presidential advisor Bernard Baruch) advocated extending nuclear knowledge to all, as long as nobody else got any nuclear weapons.

A vital component of the wartime nuclear effort orchestrated by Los Alamos was European expertise, particularly that staged from Britain. But even during the war there had been anglophobe voices, notably that of General Groves. During the war, science had outpaced security. After the war, now assisted by Gouzenko in Ottawa, breathless secret agents finally caught up. In 1950, the arrest of Fuchs in Britain spurred them into a fresh frenzy. The Fuchs case was soon underlined by the suspicious disappearance of another emigrant nuclear scientist, Bruno Pontecorvo. Born in Pisa in 1913 into a wealthy Jewish family, he became the youngest physicist in Fermi's pre-war Rome group. As well as being a notable nuclear physicist, he looked like a film star and was an accomplished sportsman who could play a fine game of tennis. It was a talented family: one brother, Gillo, became a film director, and another, Guido, a geneticist. Even before Mussolini's race laws, Bruno Pontecorvo moved to Paris in 1936 to continue nuclear physics with Frédéric Joliot-Curie, whose ardent communism later earned him the Stalin Peace Prize. While in Paris, Pontecorvo was influenced by Joliot in more than just nuclear physics. After the Nazi occupation of the French capital, Pontecorvo emigrated to the USA, without any immediate job prospect, but soon used his scientific skills in oil prospecting, before joining the British wartime nuclear research team, which had moved to the safety of remote Canada. There he learned a lot about the nascent science of nuclear reactors, going on after the war to join the UK's new research establishment at Harwell, where his recruitment, subject to a security clearance, proceeded smoothly. Unlike Fuchs, Pontecorvo had never worked at the heart of the wartime bomb effort, and was not an expert in nuclear weapons,[35] but he did know a lot about reactors.[36] While at Harwell, he was offered a job at the University of Liverpool, but before taking up the offer, he suddenly disappeared from view, after a September 1950 visit to Italy, and a subsequent trans-European journey which appeared to terminate in Helsinki.

Amid the alarm of the Fuchs case, and with the increasing pace of US and Soviet nuclear tests, Pontecorvo was immediately branded as another defector who had taken nuclear secrets to the Soviet Union. But Pontecorvo's first love was pure research, where he suggested fruitful ideas in the science of neutrinos, weightless particles produced in radioactive decay which were only discovered in 1956. In 1955, Pontecorvo came out into the open at the Soviet research establishment at Dubna, north of Moscow, where he became a figurehead of Russian subnuclear research. Towards the end of his life, humbled by Parkinson's disease, he appeared at scientific meetings outside the Soviet Union.

In 1946, all these defections and leaks had not yet become apparent, but the case of Gouzenko had made it clear that the secret nuclear effort had been penetrated. Investigations began, but even before they had borne fruit, accusations began to fly. Szilard moved too fast to become a target, while almost all of the other European exiles were back in universities, where they could do little harm. One still prominent in the ideological nuclear struggle was Eugene Rabinowitch, who had been part of James Franck's wartime team at Chicago. He was born in 1901 into an emancipated and comfortable Jewish family in St Petersburg, whose fortunes changed abruptly in 1917 with the Russian revolution. Their migration was hindered by the First World War and its aftermath, but finally they settled in Berlin, which, despite its post-war trauma, offered more stability than cities the Rabinowitches had known further east. Subsidizing his studies by writing for the Russian expatriate press, Rabinowitch progressed to a post at Göttingen under Franck. In 1933, Franck was able to find a niche for Rabinowitch with Bohr in Copenhagen. Like several of his contemporaries, he moved on to England, but soon became frustrated by the lack of prospects, and finally moved to the Massachusetts Institute of Technology in 1938. There he was recruited for nuclear development work at Chicago, and reunited with Franck.

A concerned scientist, Rabinowitch helped compile the 1945 Franck report, and was one of the founders of the *Bulletin of the Atomic Scientists*, the official mouthpiece for scientists pushing for the transfer of nuclear affairs to the civil sector. Such visibility was not always to his advantage. With the McMahon Act and the US Atomic Energy Commission still in the pipeline, the military continued to pull the strings. Rabinowitch was denied security clearance for a job at the Oak Ridge nuclear plant, built to separate uranium and plutonium for bombs. This decision was subsequently revoked, but Rabinowitch had meanwhile decided to quit nuclear physics and move instead to biophysics at the University of Illinois.

By the time the McMahon Act was passed, the source of the wartime security leaks had become clearer, while East–West relationships had chilled. Instead of the originally foreseen 'free dissemination' and 'maximum liberality', new compromise legislation set out to control information, with all material on nuclear matters being deemed secret, unless specifically indicated otherwise. In this atmosphere of doubt and mistrust, suspicion was not confined to the immediate management of nuclear affairs, and communists, real or imaginary, became ready targets. In all this meddling, the activities of the House Un-American Activities Committee (HUAC) would become the most notorious.

The effort which eventually became HUAC had begun after the First World War, originally to investigate German involvement and infiltration in American affairs. But as the memory of that war began to fade, the objective soon switched to a new perceived target. Some thought that communist objectives were not limited to the political arena. Culture could also be a target, as had happened in Nazi Germany. HUAC itself was established in 1938 and its subsequent activities sometimes resembled a medieval witch-hunt. US culture was alleged to have become contaminated by leftist propaganda. In scenes eerily reminiscent of the Nazi era a decade before, in 1947, HUAC began to investigate the Hollywood motion-picture industry. Figures such as

Ronald Reagan and Walt Disney were called to testify. Others refused, objecting to vehement HUAC attacks on unsuspecting and unwarranted targets. Hundreds were blacklisted, their careers jeopardized. In scenes which mirrored Einstein's move from Nazi Germany to the USA in 1932, in 1952, silent movie star Charlie Chaplin, an emigrant from Britain who had lived and worked in the USA since 1910, departed for Europe and was not allowed to return.

HUAC's efforts redoubled as anti-communist sentiment escalated following the outbreak of the Korean War in 1950. They soon focused on nuclear science, its reputation weakened by revelations of espionage. Edward Condon, one-time advisor to the Senate Special Committee on Atomic Energy, became a victim. Appropriately born in atom bomb country, Almagordo, New Mexico, he had studied in Göttingen and Munich before returning to the USA and becoming one of the nation's experts on quantum mechanics. In 1943, he had been recruited for a senior position at Los Alamos, but resigned after just six weeks, in protest against the strict militarization of the project, thus earning the disapprobation of General Groves. He was also a fervent internationalist, and had been a wartime member of a low-level group called the American-Soviet Science Society. After the war, he moved into a visible and influential position as Director of the US National Bureau of Standards, but quickly became a target. In March 1947, the *Washington Times-Herald* had broadcast on its front page 'Condon Duped into Sponsoring Commie-Front Outfit's Dinner'.[37]

Condon soon appeared on HUAC's ultra-sensitive radar screen. As the attacks on him gathered momentum, scientists spoke up for him. Einstein declared the accusations 'a disservice to the interests of the United States', and wartime nuclear experts, together with Einstein, held a dinner in Condon's honour.[38] Folk tales of HUAC sessions[39] report some absurd questioning redolent of the *Judenphysik* era: 'Dr. Condon, it says here that you have been at the forefront of a revolutionary movement in physics called "quantum mechanics". It strikes this hearing that if you could be at the forefront of one revolutionary movement ... you could be at the forefront of another.' In 1954, in the quest for security clearance for a new assignment, Condon's application was revoked by Vice-President Richard Nixon. Another HUAC victim was Frank Oppenheimer, brother of the former Los Alamos Director, who had also worked at Los Alamos and was dismissed from his position at the University of Minnesota. Frank Oppenheimer's history of association with communists soon made another, much larger, target emerge.

With questionable links to communist sympathizers deep in his personal history, J. Robert Oppenheimer, once a hero, also came under attack. One accusation was that he had deliberately held back Teller's idea for a hydrogen bomb so that the USA would fall behind in the subsequent arms race with the Soviets. Many of his former colleagues rushed to his aid, but they were no longer influential. However one still was. Teller, passed over several times at Los Alamos, had an axe to grind. In 1954, Oppenheimer appeared in front of a specially orchestrated court, where the prosecution, brilliant and well prepared, snared a flat-footed and over-confident Oppenheimer. His testimony was torn apart.[40] He lost his security clearance, vital if he were continue to be influential in US nuclear policy, and he emerged a broken man. But in humiliating Oppenheimer, Teller lost most of the few scientific friends he still had. Later, he attributed his statements to his 'confusion' when called to testify.[41]

As such big heads rolled, young candidates for new jobs were called to swear convoluted oaths. Applications for research grants also became subject to loyalty declarations. Tenured professors in California not only had to affirm loyalty to the state constitution, but had to deny membership or belief in organizations (including Communist ones) hostile to the United States government. Some found such demands offensive and undignified, and refused to cooperate: 31 faculty members of the University of California lost their jobs.[42] In one instance this further blighted a career which had already suffered, that of the historian Ernst Kantorowicz (1895–1963). Born in Pozen in Prussia (now Poznan in Poland) into a wealthy, assimilated Jewish family, he became controversial, if not famous, in Germany through his romantic biography of the Holy Roman Emperor Frederick II ('*Stupor Mundi*'), but in 1933 had to leave his post at Frankfurt, moving first to Oxford, and then to the University of California, Berkeley. After refusing to take the oath in 1949, Kantorowicz moved to Princeton's Institute of Advanced Studies, where ironically he was posthumously accused of Nazi sympathies.[43]

Also among the dissidents was Gian Carlo Wick, one of the few Italian scientists who had remained in Rome during the war, and Jack Steinberger, who had left Germany while still a schoolboy. They moved to the more relaxed atmosphere of the US East coast. Also affected was

David Bohm, born in the USA in 1917 into a family of Hungarian Jewish immigrants. Bohm had dabbled in radical politics and had been denied a security clearance to work at Los Alamos. He had been working on nuclear effects at Berkeley, and with no security clearance was initially not even allowed to look at his own research work. Oppenheimer intervened on Bohm's behalf, so that he could at least finish his PhD work at Berkeley.[44] After the war, Bohm moved to Princeton, where he became a colleague of Einstein. Summoned by HUAC in 1951, Bohm was acquitted, but still lost his job at Princeton, and left the USA altogether.[45]

Once, the nation had been a bright beacon for scientists. Now, movement of scientists both into and out of the USA was tightly controlled. Routine applications by US citizens for passports were refused. In 1952, two years before he won the Nobel Prize for Chemistry, Linus Pauling was one of those denied a travel document. In the ensuing years, there were some six hundred others. As well as Americans wishing to travel abroad, those applying for visas to enter the USA also came under examination. In 1952, about half the scientists applying to enter the USA experienced difficulties,[46] among them Rudolf Peierls, the man who had hired Fuchs, and Mark Oliphant, who had helped convince the USA to begin work on the Atomic Bomb. In the face of such difficulties, some scientific organizations boycotted the USA.[47]

The Cold War was an inglorious period for showcasing US science. After having been chased to the thermonuclear hydrogen bomb, in 1957 the nation was wrong-footed as the Soviets startled the world with their Sputnik satellite and follow-up exploits in space. The humiliated Americans resolved to do better, and launched a massive catch-up effort. They remembered the expertise of Wernher von Braun and his Germans colleagues who had been spirited away from their defeated nation in 'Operation Paperclip', but who had languished in the military background ever since. In 1958, von Braun re-emerged as leader of rocket development for the new National Aeronautics and Space Administration (NASA), whose Apollo effort was crowned ten years later by the dramatic moon landing. As well as von Braun, Paperclip had assembled some 50 German missile scientists, the brains behind the impressive wartime V2 rocketry, and spirited them across the Atlantic. The new US space endeavour was helped by European brains in much the same way as the nuclear programme had been in the previous decade. But while the Jewish nuclear brains had been forced to get out of Europe before the war, space brains were taken out afterwards and commandeered into service. Rockets are large, operate in three-dimensional space, and do not approach the speed of light. Rocket science had no debt to Einstein, and had not attracted Jewish intellect in the same way that quantum science had.

The 1980s was a unique era, symbolized in the West by the hard-line leaderships of US President Ronald Reagan and UK Prime Minister Margaret Thatcher. It was an extravagant decade where appearance often seemed more important than content. In finance, abstemious convention was pushed aside in a new policy of heady deficit spending, with consumers buying trendy goods on credit and financiers rushing to trade dubious 'junk bonds'. In Eastern Europe, sporadic popular uprisings traditionally had been crushed by the tracks of Soviet tanks, but in the 1980s, the Polish *Solidarnosc* movement under Lech Walesa presaged new political developments: others followed in Hungary, Czechoslovakia, and Romania. Soviet head of state Mikhail Gorbachev boldly overthrew decades of authoritarianism, introducing *perestroika* (reconstruction) and *glasnost* (openness) reforms, astonishing onlookers in the East and West alike. Amid such unpredictability, nuclear strategy had entered its own junk bond era. Until then, defence had been advertised as retaliation in kind against any nuclear attack. Retaliation was defence and defence was retaliation—the Mutual Assured Destruction scenario formulated in the von Neumann years. However, in 1983, nuclear arms policy appeared to break out of this impasse with President Reagan's announcement of a new Strategic Defense Initiative (SDI)—'I call upon the scientific community who gave us nuclear weapons to turn their great talents to the cause of mankind and world peace: to give us the means of rendering these nuclear weapons impotent and obsolete.' The grand effort was quickly rechristened 'Star Wars', after the successful 1977 George Lucas science fiction film.

The scientific expertise of the USA and the vast resources of US industry duly followed Reagan's exhortation. In a carefully scripted scenario, incoming Soviet missiles carrying any of a vast arsenal of nuclear weapons would be picked up by layered screens of satellite-borne sensors. These would launch sophisticated new devices to knock out the incoming weaponry well before it reached its objective. Soviet warheads released en route would be smothered by thick neutron shields, volatilized by laser cannon, or destroyed by 'smart rocks' or 'brilliant pebbles' launched from orbiting battle stations. Nuclear explosions would be orchestrated in space to ignite antimissile X-ray lasers, a *son et lumière* of unimaginable power, the ultimate Nintendo game.

When Reagan said, 'I call upon the scientific community who gave us nuclear weapons', most of that community had cut their connections to those weapons. One exception was Teller, now at Livermore National Laboratory, still a champion of the nuclear cause, especially when it had an unconventional slant. He had enthusiastically pushed the fusion bomb while its fission counterpart was still in the future. When the hydrogen bomb finally became a reality, Teller had urged it to be used to blast out new harbours in Alaska, or to extract oil from low-grade tar sand. In 1967, he visited Israel to advise scientists and politicians on that nation's nuclear weapons programme. He attributed his heart attack in 1979 to spirited confrontations with prominent environmental and anti-nuclear lobbyists after the Three Mile Island nuclear power station accident. Teller was an influential advocate of Star Wars. As he neared his 80th birthday, the project provided a fresh opportunity for the old warrior to put on his war paint.

During the Second World War, the science behind the fission bomb effort had been largely confined to the scientists working on it. Few others understood. With Star Wars it was different. Some influential scientists, among them Bethe, began to question the feasibility of such ideas, criticized in different quarters as either ambitious or far-fetched. The mind which had decoded the intricate gas dynamics of the first mushroom cloud, millisecond by millisecond, was also able to track the proposed Star Wars scenarios. Perhaps still mindful of what had happened to Oppenheimer, Bethe was also questioning Teller's political ability. But whatever its scientific basis, Star Wars had another rationale. It was also a brazen display of scientific chest-thumping which the Soviets had not been able to match. Where once they had been ambitious, convinced that they could match, even outdo the Americans, now they were terrified. The post-war arms race may have boosted Western supply-side economics, but it bankrupted the Soviet Union. They still had agents, but there was no Star Wars equivalent of Klaus Fuchs to keep them reliably informed. The visibility of Star Wars, underlined by images from screen science fiction, helped change the mood of the world. In the giant political poker game of the 1980s, the Soviets blinked first. Soon the concrete Berlin Wall fell, and its aftershocks brought down the rickety house of cards that had been the empire of the Soviet Union. The Cold War evaporated just as quickly as it had arrived. The USA, now by default the ultimate world power, no longer needed Star Wars as a propaganda weapon and the nebulous project was downgraded. US weapons science was now unchallenged.

During the Star Wars era, there had been some real stellar science. In February 1987, astronomers were excited by a new light in the sky. Supernovae are the ultimate in stellar explosions, when a star, its nuclear fuel spent, collapses under the gravitational crush of its own mass. About twenty are seen in the depths of space through telescopes each year, but those near enough to be visible to the naked eye are rare. Before 1987, they had been recorded in 1006, 1054, 1572, and 1604. But in 1987 they were also monitored by new devices, the neutrino detectors first advocated by Bruno Pontecorvo. The detailed diary of the various signals provided new insight for scientists studying stellar mechanisms. Prominent among them was Bethe, aged 86. Half a century before, he had shown what makes the Sun shine. A master of understanding complex mechanisms, he was now looking at what happened when a star died. Since the late 1930s, when he and Teller had worked together on the theory of nuclear forces, their interests had diverged. While Bethe had long since left the weapons arena, Teller was the last of the nuclear Mohicans from the 1930s exodus. Many saw in him elements of the demented Presidential scientific advisor of Stanley Kubrick's 1964 black comedy *Dr Strangelove: How I Learned to Stop Worrying and Love the Bomb.* Misunderstood and vilified, but a fine scientist, Teller died in 2003, aged 95.

NOTES

1. Rhodes MAB, p.374.
2. Rhodes DS, p.247.
3. Poundstone, pp. 33–4.
4. Hargittai, p.30.
5. Felix Bloch, AIP oral history, 4509.
6. Von Neumann, J., Morgenstern, O., Kuhn, H.W., andRubinstein, A., *Theory of Games and Economic Behavior*, Princeton University Press, reissued 2007.
7. Poundstone, p.96.
8. The colleague was Jacob Bronowski, who related the episode in his TV series, *The Ascent of Man* (BBC 1973).

9. Poundstone, p.180.
10. Ulam (Appendix 2).
11. von Ardenne, M., *Ein glückliches Leben für Technik und Forschung*. Verlag der Nation, Berlin, 1982, p.389.
12. Rhodes DS, p.162.
13. Rhodes MAB, p.759.
14. Rhodes DS, p.209.
15. Jungk, p.249.
16. Rhodes DS, p.356.
17. Rhodes DS, p.418.
18. Rhodes DS p.455.
19. Rhodes DS, p.466.
20. Rhodes DS, p.471.
21. Pais ELH, p.230.
22. Rhodes MAB, p.305.
23. Pais SL, p.8.
24. Pais ELH, p.129.
25. Pais ELH, p.118.
26. Pais SL,
27. 1940 Conference at the Jewish Theological Society of America, quoted in Pais ELH, p.122.
28. Many of these criticisms are cited in Dawkins, R., *The God Delusion*, Bantam, London, 2006, pp. 37–8.
29. Jammer, M., *Einstein and Religion*, Princeton University Press, 2002, p.104, quoted in Dawkins (Note 28), p.37.
30. Pais ELH, p.233.
31. Pais ELH, p.238.
32. Pais ELH, p.236.
33. Pais ELH, p.230.
34. Jerome, F., *The Einstein File, J Edgar Hoover's War Against the World's Most Famous Scientist*, St Martin's Griffin, NY, 2003.
35. Fermi, p.257.
36. Rhodes DS, p.129.
37. *Washington Times-Herald,* March 23 1947, quoted in Wang, p.131.
38. Wang, p.139.
39. Sagan, C., *The Demon-Haunted World, Science as a Candle in the Dark*, Ballantine Books, New York, 1997, p.248.
40. Rhodes DS, p.544.
41. Teller, p.383.
42. http://sunsite.berkeley.edu/~ucalhist/archives_exhibits/loyaltyoath/
43. Cantor, N.F., "The Nazi Twins: Percy Ernst Schramm and Ernst Hartwig Kantorowicz," In *Inventing the Middle Ages,* Harper, New York, 1991.

44. Peat, F.D., *Infinite Potential, the Life and Times of David Bohr*, Addison Wesley, Reading, MA, 1996, p.64.
45. Peat (note 44), p.120.
46. Wang, p.278.
47. Wang, p.279.

Epilogue: Europe redux

After the mass departure of the wartime scientists from Los Alamos, the place lost a lot of its momentum and sense of purpose.[1] As the ongoing effort slowed to a rhythm that was only a faint echo of its former throbbing pulse, the military nevertheless realized that it might have to be reactivated at any time. To safeguard its interests, the nation should be able to galvanize a similar effort in the future, just as it had done in 1942.

First, some remedial action was needed: the war had siphoned off potential university students, and to restore the nation's supply of graduates and other trained specialists, the 1944 GI Bill (Servicemen's Readjustment Act) entitled demobilized soldiers to a college education. Many jumped at the opportunity. But it was not just a matter of restoring the flow of science students through the universities. The war had changed the whole scale of doing research. Before 1941, US government funding for university science had been meagre. In 1946, the newly established US Atomic Energy Commission prepared to take up the reins from the wartime administration of nuclear affairs. The military, which had nursed the wartime effort, was reluctant to let its still immature nuclear child leave its protective nest.[2] The War Department (and the Navy, then separate) thus spent lavishly on basic research, leading to a boom in post-war nuclear physics. (In 1947, the War Department changed its name to the less bellicose Defense Department, which included the Navy.) In the post-war decade, Defense Department spending on scientific research and development jumped from 245 million to 1.5 billion dollars.[3] Some scientists naturally welcomed such generosity. This largesse even found its way abroad, where university departments had extra administrative costs covered, or qualified for US military transport flights. If this came with strings attached, they were not always visible.

The wartime bomb effort had built vast new laboratories and plants on undeveloped sites at Los Alamos, Oak Ridge, and Hanford. To nurture the nascent post-war effort, existing university laboratories, such as those at Berkeley and Columbia, were upgraded, and new ones planned. The wartime effort at Chicago University had led to a new research centre at the Argonne Forest south-west of the city.

Argonne became the first of a series of new National Laboratories. In 1945, Vannevar Bush's visionary presidential report, *Science, the Endless Frontier,* stressed the role of science as a the pacemaker for technological progress, and led to the establishment of the US National Science Foundation to oversee and coordinate academic research in science and engineering, and to divert universities' attention from the military munificence. As Roosevelt's scientific advisor, Bush had been influential in creating a military-industrial science base which was in danger of overshadowing the true nature of scientific research. The report underlined how the USA had been caught unawares by the war and had to run fast to catch up. It also amplified the voices in US science which pointed to the risks of pure research at many universities being underwritten by military money. 'Many of the lessons learned in the war-time application of science under Government can be profitably applied in peace. The Government is peculiarly fitted to perform certain functions, such as the coordination and support of broad programs on problems of great national importance. But we must proceed with caution in carrying over the methods which work in wartime to the very different conditions of peace. We must remove the rigid controls which we have had to impose, and recover freedom of inquiry and that healthy competitive scientific spirit so necessary for expansion of the frontiers of scientific knowledge'.[4] (It was a theme which would recur. In a memorable farewell address in January 1961, President Dwight D. Eisenhower warned of the dangers of the increasing influence of powerful contractors and influential decision takers in what he called the 'military-industrial complex'.)

Some frontiers were already beckoning. As well as producing bombs, the wartime programme had led to the first nuclear reactors, prototypes for a bright new age of energy, as yet unclouded by concerns about radioactive waste and the environment. The effort had also introduced another new range of scientific engines. In 1929, Ernest Lawrence in Berkeley had developed a new device—the cyclotron—to produce particle beams to probe the tiny nucleus deep inside the atom. The same misunderstanding which led to explosive devices exploiting nuclear fission to be called 'atomic bombs' also named these new research machines, 'atom smashers'. Smashing atoms is literally child's play. Lawrence's machines smashed nuclei. During the war, these techniques were used in huge new nuclear grindstones—'calutrons' or 'racetracks'—which whirled uranium vapour around a huge electromagnet and winnowed out the lighter uranium 235. With a wartime shortage of copper for the windings of the electromagnets, the US Treasury gallantly lent the project 13,000 tons of silver bullion to be recast into electrical wire. Each microgram of the 60 kilograms of uranium 235 used in the Hiroshima bomb had been separated out by these magnets.[5]

As cyclotrons were made bigger and became more expensive, they ran up against a problem. As they hurled their nuclear particles faster and faster, these speeds began to enter the domain of Einsteinian relativity. Tiny shifts in mass and velocity made the particles whizzing around the machines fall out of step with the electromagnetic kicks which were supposed to accelerate them on successive orbits, and the beams stalled. To overcome this, more new machines appeared—synchrocyclotrons and synchrotrons—which could take relativity in their stride. In the 1930s, Lawrence's cyclotrons had already loosened the European stranglehold on nuclear physics research. The new synchrocyclotrons and synchrotrons made it an all-American affair. Where Rutherford had once changed our picture of the world with home-made table-top gadgets, laboratories now required huge, expensive machines. Intrepid European scientists had carted heavy apparatus to the tops of mountains to catch rare subnuclear gems from outer space, before they were swallowed up by the atmosphere: American scientists could now manufacture their own subnuclear jewellery. In 1950, a Berkeley synchrotron provided the first new particle to be discovered in this way. The experimental team included two men who had left Germany in 1934, when they were still school students: Jack Steinberger,[6] who after his war service had learned science through the GI Bill and had later worked under Fermi and Teller in Chicago, together with Wolfgang Panofsky, born in Berlin in 1919, who had emigrated with his father, an influential art historian. With money flowing freely, new 'engines of discovery'[7] attracted eager young researchers and quickly became the pacemakers for a post-war renaissance of subnuclear physics.

But, just as Lawrence's huge wartime racetrack had warranted a specially built home, and Fermi's nuclear pile had outgrown the facilities at Columbia, so these cyclotrons and their lookalikes soon became too big for most university departments. The newly built US National Laboratories were equipped with gleaming new 'atom-smashing' machines whose dimensions rivalled those of Lawrence's wartime racetracks. These could synthesize subnuclear matter which before could only be dredged from the cosmic wisps that arrived on Earth from outer space. A vast new scientific complex, the Brookhaven National Laboratory, was built on a former army base on New York's Long Island. Initially under the jurisdiction of the US War Department, it was transferred to the Atomic Energy Commission in 1947. Brookhaven's first synchrotron was the 'Cosmotron', thus called to underline its ability to fake extraterrestrial matter. In 1948, the AEC called for the construction of two even bigger synchrotrons, one at Brookhaven, the other (the Bevatron) at Berkeley. As these giant machines and others got into their stride, discoveries duly followed, which were followed in turn by invitations to visit Stockholm. In the first half of the twentieth century, just six US scientists had been awarded the Nobel Physics Prize: in the second half there were ten times that number.

This simple statistic underlined how, across the Atlantic, Europe and European science had been ruined by the war. The Old Continent had also been split in two, with Western Europe striving hard to emerge from the ruins of the war. Further East, Germany had been bisected, and lands which once had been close to the centre of science were now isolated. Embarrassed by having been the scene of two conflicts which were then inflicted on the rest of the planet, Western Europe resolved to present a united front to the rest of the world. These whispers eventually became today's European Community. But for science, there was an even more pressing need. European scientists could only look on ruefully at the new US predominance in a field they had invented only a few decades earlier. The forced migration of scientists from Nazi Europe in the 1930s had now grown into a tradition where each new generation of European quantum talent automatically looked across the Atlantic for its scientific fulfilment. The Quantum Exodus had been replaced by the 'Brain Drain'.[8] Only the wealth of the USA looked capable of supporting the giant new installations needed to do 'Big Science'. Could this be reversed?

There had been a brief post-war subnuclear coda in Britain. At the new Harwell research establishment, experts from Chadwick's cyclotron at Liverpool and from the UK's wartime radar development programme built a synchrocyclotron. The buoyant Oliphant in Birmingham pushed for an even bigger synchrotron. But post-war British austerity quickly strangled such initiatives. This underlined that the huge installations springing up across the USA were beyond the means of individual European nations. To provide a new stage for European science, ideas emerged for a pan-European (or at least Western European) centre to provide resources on a continental, rather than a national, scale. Supporting voices came from many directions, particularly from Edoardo Amaldi (1908–89) in Rome, who had been left scientifically stranded when Fermi and his colleagues had fled following Mussolini's 1938 racial dictates, and had watched his nation's nuclear physics expertise vanish.[9] Niels Bohr, back in Copenhagen, was another important voice. Thus came into being in 1954 *Le Conseil Européen pour la Recherche Nucléaire*, the European Council for Nuclear Research, or CERN. The acronym was retained despite the official name subsequently being changed to 'Organization' rather than 'Council'[10]—CERN slips off the tongue more easily than OERN. From a green-field site near Geneva, Switzerland, its ambitious mission statement was to provide facilities for European subnuclear research to match those of the USA.

Having established the new organization, the search began for someone to head it. At Bohr's suggestion, attention soon focused on Felix Bloch in Stanford, who had just been awarded the 1952 Nobel Physics Prize for his work in developing new methods of measuring nuclear magnetism. Bloch's name was well known: he had worked with Heisenberg (in fact he had been Heisenberg's first doctoral student), and later had been selected by the astute Pauli as a research assistant; physics students learn about 'Bloch waves' of quantum electrons in solids, and 'Bloch walls' in magnetism, work that he had done even before leaving Germany in 1933. On returning to his study of nuclear magnetism after the war, Bloch put his microwave knowledge to work. After Otto Stern had opened the way in the 1920s, Isidor Rabi and others had shown how to study nuclear magnetism by passing nuclear beams near a powerful magnet. Bloch found that if a static nuclear sample is used instead and put into a strong enough magnetic field, some nuclei shift position and line up in the direction of the field. If a carefully oriented high-frequency signal is sent through the sample, the nuclei begin to twist, sending out their own radio call-signs. At first, this was just another way of studying nuclear magnetism, avoiding the difficulty of having to shape nuclei into beams, but careful work extended it into the new technique of nuclear magnetic resonance imaging. This revolutionized medical diagnostics and earned, for Paul Lauterbur of Illinois and Peter Mansfield of Nottingham, the 2003 Nobel Prize for Medicine.

But that was far in the future in 1954, when Bloch was invited to become the first Director General of the new European laboratory, construction work for which was just beginning. The return to Europe of a scientist who had chosen exile in 1933 was also highly symbolic (even though in the meantime Bloch had taken US nationality). However, Bloch himself had strong reservations: with his own research on nuclear magnetism still productive, he did not want to abandon it and move into scientific administration, even at such a prestigious level. He stipulated to CERN that he would only stay for two years: he would not formally give up his Stanford post, and would remain on leave of absence. He also wanted his laboratory equipment and some assistants to move to Geneva.[11] Once installed at CERN, Bloch carried on with this research and delegated much official business to Amaldi, who had not been interested in the directorship himself, but was now burdened with it. When Amaldi returned to Italy, Bloch turned to the head of CERN administration. Bloch thought that the daily business of running CERN could be completed in less than an hour. With the conflict of interest growing, in 1955, Bloch informed CERN that he was not prepared to extend his stay beyond the autumn of that year. He meanwhile also exercised an option with Bohr in which they had agreed that Bloch could take his final decision even after arriving at CERN. He soon returned to Stanford, leaving the infant laboratory leaderless.

As CERN struggled to find its feet, Cornelis Bakker, a Dutchman who stepped into the directorship left vacant by Bloch's departure, was killed as his aeroplane crashed on approach to New York. Two heavyweight European physicists were headhunted as potential CERN leaders: Hendrik Casimir at the Philips electronics concern in the Netherlands, and Victor Weisskopf, then at the Massachusetts Institute of Technology. Before the war, both had been assistants to Pauli in Zurich. However Casimir quickly said that he was not interested in leaving his influential position at the Netherlands electronic giant. On the other hand Weisskopf's stature as a scientist and a humanist, and his language abilities (German, English, and French) made him well suited for the job, and he relished the idea of returning to the Old Continent.

Now an international figure, in 1961 Weisskopf returned to Switzerland. Apparently, the country had forgotten what had happened in Zurich a quarter of a century earlier when he had been accused of being involved in communist affairs.[12] When he reminded the authorities of their earlier letter, they 'laughed it off'.[13] Once at CERN, Weisskopf's charisma immediately charmed the disoriented laboratory: his initial two-year mandate was extended until 1965. Before his arrival, CERN had been split between two objectives. An ambitious team, including key members of the squad that had built the Harwell machine, was making excellent progress on its first large synchrotron, but what was it going to be used for? The objectives of the laboratory's future research programme were far less clear.[14] The scientists wanting to carry out research at CERN still saw themselves primarily as British, French, German ... To remedy this, Weisskopf encouraged them to join together in international collaborations. One of the much-appreciated features of his leadership was his habit of dropping in unannounced on scientists at work, a technique he had probably picked up from Oppenheimer at Los Alamos. Many detailed decisions had to be made during Weisskopf's time at CERN, but perhaps his biggest contribution was to its style. European institutions can become staid and cumbersome, but Weisskopf ensured that CERN could count on the smooth cooperation of scientists, specialists, and administrators from a wide variety of backgrounds and cultures, ensuring that they worked together harmoniously in a team whose combined strength ultimately became bigger than just the sum of its different parts.

Most of Weisskopf's Los Alamos contemporaries had by now retreated into the academic background. There was a notable exception. Totally disillusioned with nuclear physics, Leo Szilard left the field completely after 1945, and despite approaching the age of 60, turned his attention to a totally new scientific topic. Instead of the nuclei at the heart of atoms, he focused instead on the large molecules of living organisms, the new science of molecular biology. With the objective now to understand the genetic basis of life, he was soon collaborating with another European emigrant and former quantum physicist, Max Delbrück. Szilard's inbuilt predictor, still working well, told him that this science had an important future. Again he was right: but he had not been the first to see. One important portent had come after the death of Rutherford in 1937. Instead of appointing one of Rutherford's lieutenants, themselves now Nobel laureates and famous, as his successor, the Cambridge authorities judged that, after leading the world in subatomic physics for half a century, it was time for the university's Cavendish Laboratory to change course. Their chosen new direction was crystallography, using beams of X-rays to illuminate the large molecules. The reward for this foresight duly came in 1953 when Francis Crick and James Watson revealed the double-helix structure of DNA, 'the secret of life'. It was the overture for a new age of science. Szilard now looked to see what direction this new science should take. The research that had paved the way for molecular biology had been done in Europe. The Old Continent had developed these vital X-ray techniques, and it should now consolidate this tradition and advance the science it had nurtured. Otherwise the Americans would step in, as they had done for nuclear physics. Szilard's 1961 book *The Voice of the Dolphins* envisaged the creation of an international European biology research laboratory to nurture this new science, just as CERN was setting out to do in another field.

During this time, Szilard was diagnosed with bladder cancer, and devised his own programme of radiation therapy. As the cancer regressed, Szilard's predictor began buzzing again, but this time for a different reason. With the USA and the Soviet Union staring provocatively at each other, Soviet missiles had been spotted on Cuba. The Cold War suddenly threatened to become extremely hot, and in 1962 Szilard decided it was time to be one jump ahead of another war. He rushed over to Europe, where the first door he knocked on was that of his former Los Alamos colleague, Weisskopf, now Director General of CERN. Weisskopf relates Szilard's unexpected appearance: 'World War III is about to break out,' Szilard announced. 'I am the first refugee from America. I have … brought all my belongings'.[15] As well as a desk at CERN, Szilard demanded to talk to Soviet leader Nikita Khrushchev. However, the international crisis was soon defused without Szilard's help, and he turned his attention back to molecular biology. Also passing through Geneva at the time were James Watson and John Kendrew, en route to Stockholm to receive their Nobel awards for the elucidation of molecular structure. Watson's award (with Francis Crick and Maurice Wilkins) was for Medicine and Physiology, for their discoveries concerning the molecular structure of nucleic acids and its significance for information transfer in living material, while Kendrew's Chemistry award (with Max Perutz) acknowledged their discovery of the structure of certain proteins. Both of these developments had been the fruit of X-ray experiments at Cambridge, underlining the wisdom of deciding on that new research path. But Cambridge could not do scientific research for an entire continent, and Szilard suggested to the two new Nobel laureates that they set up a new European Molecular Biology Laboratory, modelled along CERN lines, to extend the work carried out in Britain and to parallel the by-now considerable effort underway in the USA. The laboratory duly came into being in 1974.

However, in spite of the impetus and spirit injected by Weisskopf, CERN nevertheless remained handicapped by European inertia and unwieldiness. This encouraged paying attention to detail, but prevented it from sharpshooting at targets of research opportunity as soon as they popped up, an activity at which the pragmatic Americans excelled. For some two decades, the European laboratory had to content itself with playing catch-up. However, with alpine skiing and French wine country close at hand, and ideally sited in the centre of Western Europe, CERN provided an attractive European haven for travelling scientists, and for sabbatical researchers seeking a cosmopolitan atmosphere. It inherited the role that Bohr's institute in Copenhagen had played prior to 1939. It had its own impressive atom-smasher, but at first little startling science emerged from its fussy research programme.

At least in the West, the USA appeared to have a monopoly on cyclotrons and their derivatives,[16] which had become a US speciality. Like another all-American activity, baseball, attempts by others to take part were looked upon patronizingly and not always taken seriously. CERN itself was not entirely blameless: while Brookhaven had chosen to label its first major machine as the 'Cosmotron', CERN's scientists were content to call theirs the uninspired generic 'Proton Synchrotron'. Nobody bothered to name the ship before it was launched. Bereft of vision, its research programme stumbled forward. But slowly things began to change. In the 1970s, both the USA and CERN commissioned gleaming new synchrotrons to hurl beams of protons into a still higher energy range. The objective now had switched from the interior of the nucleus to the inside of its constituent protons and neutrons, to seek signs of the enigmatically named 'quarks' inside. Even though CERN had dared to call its new machine the 'Super Proton Synchrotron' (SPS), it was initially handicapped. A superlative name was not enough. Flamboyant leadership ensured that a machine at the new US Fermi National Laboratory (Fermilab) near Chicago had the edge in energy, and was commissioned earlier than expected. Later to be called the Tevatron, it looked set to continue the US domination of this science.

In 1955, Emilio Segrè and Owen Chamberlain, using the Berkeley Bevatron, at the time the world's biggest synchrotron, had discovered the antiproton, the antimatter equivalent of the proton. It brought Segrè and Chamberlain the 1969 Nobel Physics Prize, and was the first subnuclear example of an eerie quantum mirror world, that of antimatter. The two mutually opposite existences of matter and antimatter had been wrenched apart by the primeval Big Bang, but the antimatter component has become so remote from our Universe as to become almost invisible. It is in the nature of science that one day's discovery becomes the next day's tool: so it had been with cathode rays, and later with X-ray crystallography. So it would be at CERN, which towards the end of the 1970s made the bold decision to adapt its new SPS, which had hardly got into its research stride, for a totally new role. Instead of hurling around protons and then releasing them to bombard remote targets, the objective now was to hurl around separate beams of protons and antiprotons. With opposite electric charges, the particles and their antimatter counterparts can whirl round a racetrack in opposite directions in carefully separated orbits, until finally being nudged together. It was a totally new way of doing science.

The Italian physicist, Carlo Rubbia, a latter-day Leo Szilard, had doggedly pushed the proton–antiproton collision idea. At first, he had brought his scheme to the Americans, who scorned it, preferring instead the traditional goal of cranking up protons to higher energy. He had then offered it to CERN. Realizing that for once the Americans had missed an opportunity, CERN stopped what it was doing with its SPS, carried out difficult proof-of-principle trials, and set off in a fresh direction. In 1984, Rubbia and the Dutch scientist Simon van der Meer, who had invented techniques to enable CERN's synchrotrons to handle antiprotons, shared the Nobel Physics Prize. US scientists, taken by surprise, were piqued. Their monopoly of the synchrotron field had been broken, and a new generation of upstart Europeans had snatched a Nobel Prize from under their noses. Their knee-jerk reaction was similar to what happened after they had been humiliated by the launch of the Soviet Sputnik satellite in 1957. Just as a shame-faced USA had then ambitiously set out to become the undisputed world leader in space exploration, so in the mid-1980s its aim was to regain its lost synchrotron prestige. Plans were rapidly drawn up for a giant new machine, the grandly named Superconducting Supercollider (SSC), a vast 87-kilometre underground ring to be built at another new national laboratory, this time near Dallas in Texas. Its scale harked back to the transcontinental immensity of the Manhattan project, something which even pan-European resources would not be able to match.

The tide turns

In the 1980s, this SSC carried the arriviste spirit of deficit spending and 'Star Wars'. However it was only scheduled to become complete in the 1990s, a decade that quickly looked very different from its predecessor. With junk bonds no longer reputable, and with federal funds evaporating in a new era of cost-consciousness, any grand scheme was in a precarious situation. With its proponents from the Los Alamos generation either dead or no longer influential, the SSC's ballooning price tag became more visible than its scientific objectives. After acrimonious face-offs in Washington, in 1993 President Bill Clinton signed the bill which killed it. The flagship of the US bid for subnuclear supremacy in the 21st century had been sunk, scuttled in its home port. Across the Atlantic, CERN suddenly found itself alone at the front of the subnuclear stage. In the shadow of the SSC, the European laboratory had meanwhile prepared plans for a more modest scheme, its obscurely named Large Hadron Collider,[17] the LHC. 'Only' 27 kilometres around, and dwarfed by its putative US competitor, the LHC had resolutely planned its own future. But while the SSC would have been built on a green-field site, with every item having to be brought in or built from scratch, the LHC had the advantage of being an ongoing project, bolted on to infrastructure which had been extended by a 27-kilometre tunnel.[18]

For thirty years, any US–CERN research rivalry had been outweighed by mutual respect and understanding. It was an American, Isidor Rabi, who had first aired the idea which was to become CERN, and which Amaldi and other pioneers had taken to fruition. Thus in Europe there was commiseration, not jubilation, at the decision to cancel the SSC. American scientists who had been building the SSC and preparing to use it now scrambled for places aboard experiments at the LHC, where they were made welcome. But they were not the first scientists to look eastwards across the Atlantic. The examples of Bloch and Weisskopf had shown how the European physics venture of CERN was anxious to re-establish contact with its past. However, neither Bloch nor Weisskopf stayed on in Europe afterwards. Bloch was anxious to get back to his research at Stanford, while Weisskopf, his duty to CERN done, returned to the Boston area. His wife insisted on going back,[19] but the couple had been partially seduced by Europe, and after leaving CERN, maintained a house in a quiet French village just across the border from Geneva. Using this as his base, for many years, Weisskopf returned to CERN each summer to give memorable lectures which inspired many generations of European researchers and students.

A few of the 1930s exiles had also made their private homecomings, not necessarily transatlantic. After retiring from his professorial post in Edinburgh in 1954, Max Born went to live quietly in Germany. More of a surprise had been Wolfgang Pauli's decision in 1946 to leave Princeton's Institute for Advanced Study and return to Zurich. After the award of his Nobel Prize in 1945, he was granted US citizenship and received prestigious offers from US universities. But wary of the military shadow looming over nuclear physics, he turned his back on the USA and returned to the post in Zurich that he had left in 1940, and was soon granted the Swiss nationality which had been denied him before the war. There, Pauli became an icon of European science, a reminder of a once-proud past. In 1951, Hermann Weyl also chose to leave Princeton and return to Zurich.

In 1961, Jack Steinberger helped carry out an experiment at Brookhaven that would later earn him and his colleagues the 1988 Nobel Prize for Physics. It showed that neutrino particles—phenomenally difficult to see anyway—also come in different kinds, something which had been predicted several years before by Bruno Pontecorvo. Before this, in 1957, Steinberger had revisited Europe for a sabbatical year during his residence at Columbia University, New York. In 1965, a further sabbatical trip took him to CERN, where he embarked on a complicated experiment to explore the properties of K mesons, among the most enigmatic of subnuclear particles. Completing the experiment demanded an additional year, and began to loosen his ties to Columbia. At the same time, Steinberger was moving from one marriage to another, and a job offer from CERN in 1968 was timely. He also relished the idea of carrying out 'research in residence',[20] rather than continually commuting between a teaching classroom and the laboratory. It was a good time to come to CERN. The infant laboratory was finding its feet, and was benefiting from new instrumentation developed by Steinberger's French next-door neighbour, Georges Charpak (1924–2010). Born in Poland, Charpak had moved with his family to France as a child. When the Germans invaded, he was captured as a member of the French resistance and sent to Dachau concentration camp. He did not look Jewish, so came out alive. He went on to earn the 1992 Nobel Prize for Physics for his innovative electronics techniques,[21] which were quickly capitalized on when Steinberger became one of CERN's Research Directors in 1969. Steinberger has remained at CERN ever since, frequently as the head of major collaborations carrying out important experiments, and as a figurehead of transatlantic science.

As US researchers regrouped around the LHC flag, the machine itself sedately pushed ahead, no longer having to match the frantic initial pace that had been set across the Atlantic. Although only a fraction of the size of its ill-starred rival in the USA, to say that the LHC was 'technologically challenging' is an understatement. Everything about it is Large. To economize on electricity, the LHC protons are guided around a 27-kilometre ring by 'superconducting' electromagnets. Superconductivity had been discovered as a scientific curiosity early in the twentieth century when researchers attempted to liquefy helium gas, the cryogenic equivalent of climbing Mount Everest. Having achieved their goal, the researchers were rewarded with unexpected discoveries. At these temperatures, helium is not just fluid, but superfluid, a bizarre material. With no viscosity, once stirred, it keeps moving; it defies gravity, flowing upwards, leaking out of any unsealed container; it penetrates through even microscopic holes. At these temperatures, some substances reshuffle their inner structure and lose all electrical resistance, becoming superconducting. A current in a superconductor just keeps going. The LHC is cooled by superfluid helium to almost one degree lower than the frigidity of outer space, making the LHC the coldest site in the Universe.[22] Anyone who has tried their hand at do-it-yourself plumbing with mere water will appreciate how difficult it is to prevent a 27-kilometre ring full of superfluid helium from leaking.

CERN had never really attracted much media interest. The 'nuclear' in its title had bemused local Geneva inhabitants, who thought CERN was high up on any Soviet list of ballistic missile targets, and confused other scientists. But most researchers working at CERN did not care about publicity: they were more concerned with papers in learned journals, and often scoffed at the well-meaning but scientifically inaccurate reports of their work in newspapers and on TV. CERN had once been a scientific Cinderella among US ugly sisters, but with the international stage to itself, the collective realization grew that the LHC was to be the biggest scientific machine ever built, with an ambition to match. CERN suddenly caught the public imagination and found itself the centre of the world's interest. Why not? The LHC aims to solve the ultimate mystery—why that matter is there at all. Led by the BBC, which had its 'Big Bang Day' on 10 September 2008, TV teams and media correspondents from major outlets across the world descended on the laboratory to watch what would happen when the button was pressed.

While some people were curious, others were frightened. Each time mankind tinkers with a new aspect of Nature, there are Jeremiahs who claim the world will end. For the fission bomb, they were worried in case a nuclear spark would ignite the whole of the Earth's atmosphere. Bethe's cool understanding had attempted to convince them,[23] but in the end it was left to the test explosion in New Mexico to provide ultimate proof. For the fusion bomb, such worries were amplified. As well as more power being released, even Teller was worried that the nitrogen of the atmosphere, chemically inert, could instead become thermonuclear fuel and add to the mayhem, destroying the world.[24] For the LHC, one perceived threat was unforeseen kinds of nuclear matter which would gobble up everything in the vicinity. Another was that the intense concentrations of energy in the LHC could produce 'black holes' and guarantee self-destruction by gravity. Court injunctions attempting to stop the LHC were served via a US District Court in Honolulu, Hawaii.[25] But the real danger lurked in an unexpected corner. Just nine days after the LHC handled its first beams, and the Jeremiahs had crept away, a defective electrical connection deep inside the machine blocked its huge superconducting electrical current, releasing a blast of energy which severely damaged about 750 metres of the ring, with confetti-like fragments spread along 3 kilometres.[26] It took a year to clear up the mess, reconfigure the machine, and restart it.

Some 1600 US scientists now participate in CERN's work,[27] so that a non-European country paradoxically provides the largest single national contingent working at the ostensibly European centre. When CERN was established in 1954, it had 12 European Member States. By 2011, this had grown to 20, with others eagerly waiting to be allowed in. But to accommodate the worldwide attraction of the 'European' centre, CERN decided to set up a new category of Associate Membership to give extra-European nations more say and more responsibility. The pendulum which had swung one way when Albert Einstein arrived in the USA in 1933 has now swung back.

This new scientific demography shows itself in a very different way. In the 1930s, people crossed the Atlantic in ships. Later, they flew in aeroplanes. Now they often do not have to go anywhere at all. During the Cold War, when the Soviets had played its Sputnik and other space trump cards, an integral part of the initial US scientific response had been the establishment of a national network to link all its large computers. As an afterthought, this large network had somehow to include seismic sensors in Norway listening for signals of Soviet underground nuclear explosions. To link back to the USA, the Norwegian signals had to pass through the UK.[28] Thus was Europe introduced to what was later to become the Internet. CERN's main business is doing experiments, but these experiments produce a massive volume of data, which has to be crunched and digested. To handle all this data and shunt it around, computer specialists at CERN made their laboratory into a major communications player, Europe's largest internet 'node'. An occasional visitor to CERN in the 1980s was a young British computer specialist, Tim Berners-Lee. In 1989, after having been asked to help upgrade CERN's management information systems for the LHC era, he had an idea. However, with CERN so committed to the LHC, Berners-Lee's brainchild got little official support. The foundling instead became public property, and Berners-Lee had to find a new job. While the LHC marched slowly towards completion, what Berners-Lee had called the World Wide Web raced ahead. The dot-com world of the 1990s was a distant internet echo of the Quantum Exodus.

NOTES

1. Jungk, p.219.
2. Geiger, R., Science, Universities and National Defense, 1945-70, *Osiris,* Vol 7 (1992) pp.26–48.
3. Jungk, p.230, citing *Business Week*, 12 January 1957.
4. http://www.nsf.gov/od/lpa/nsf50/vbush1945.htm

5. Rhodes MAB, p.601.
6. see Chapter 7.
7. Sessler, A. and Wilson, E., *Engines of Discovery*, World Scientific, Singapore, 2008.
8. The term had been invented in 1952 at the British Royal Society.
9. AIP history http://www.aip.org/history/ohilist/4485.html
10. For the pre-history of CERN, see Fraser, G., *The Quark Machines*, IOP Publishing, Bristol, 1997.
11. Hermann, R., *et al.*, *History of CERN Vol 1, Launching the Organization*, p.272, and CERN's Felix Bloch obituary publication: CERN 1984-007.
12. Chapter 10.
13. Weisskopf, cited also in Enz, p.283.
14. Hine, M., *Working with Viki*, CERN Courier, December 2002, p.12.
15. Weisskopf, p.248.
16. The Soviets had made important progress.
17. 'Hadron' is the generic name awarded by particle physicists to all particles interacting through the strong nuclear force that holds nuclei together.
18. To house LEP – the 'Large Electron Positron' collider, decommissioned in 2000.
19. Weisskopf, p.251.
20. Steinberger, p.106.
21. Charpak, G., *et al.*, *Nuclear Instruments and Methods*, Vol 80, 13, 1970, and Charpak, p.157.
22. Lower temperatures are reached in small-scale experiments to study 'Bose–Einstein condensation'.
23. Rhodes MAB, p.419.
24. Rhodes DS, p.254.
25. Dennis Overbye, New York Times, 29 March 2009.
26. Rossi, L., *The subtle side of superconductivity*, CERN Courier September 2010, p.27.
27. http://external-relations.web.cern.ch/External-Relations/obs/usa.html
28. Gillies, J. and Cailliau, R., *How the Web was Born*, Oxford University Press, Oxford, 2000.

Appendix 1: A list of emigrant scientists

This is an alphabetical list of the emigrant scientists appearing in the book—mainly physicists and mathematicians—together with a few others mentioned in passing. It is not a list of all the influential figures who left Nazi-dominated Europe, nor does it include those who lost their university jobs but managed to stay on. For a more complete list of emigrant scientists, see *Emigration. Deutsche Wissenschaftler nach 1933. Entlassung und Vertreibung*, Strauss H.A. *et al.* eds, reissued by Technische Universität, Berlin,1987. More details are in the archives of the Society for the Protection of Science and Learning at the Bodleian Library, Oxford: http://www.rsl.ox.ac.uk/dept/scwmss/wmss/online/modern/spsl/spsl.html

Arendt, Hannah, political scientist, born Hanover, Germany, 1906, died New York, 1974. Left France in 1941.

Bargmann, Valentin, theoretical physicist, born Berlin 1908, died Princeton, NJ, 1989. Left Germany in 1933.

Bemporad, Azeglio, mathematician and astronomer, born Siena, Italy, 1875, died Naples, 1945. Left Italy in 1938.

Bemporad, Giulio, mathematician and astronomer, born Florence, Italy, 1888, died Rome, 1945. Left Italy in 1938.

Bergmann, Max, chemist, born Fürth, Germany, 1886, died New York, 1944. Left Germany in 1933.

Bernays, Paul, mathematician, born London 1888, died Zurich 1977, left Germany in 1933.

Bethe, Hans, physicist, born Strassburg, Germany, 1906, died Ithaca, New York, 2005. Nobel Prize for Physics, 1967. Left Germany in 1933.

Biel, Erwin, climatologist, born Vienna 1899, left Germany in 1933, left Austria in 1938.

Blau, Marietta, physicist, born Vienna 1894, died Vienna 1970, left Austria in 1938.

Bloch, Felix, physicist, born Zurich, Switzerland, 1905, died Zurich, 1983. Nobel Prize for Physics, 1952. Left Germany in 1934.

Bloch, Konrad, born Neisse, Germany, 1912, died Lexington, Massachusetts, 2000. Nobel Prize for Physiology and Medicine, 1964. Left Germany in 1934.

Bohr, Niels, theoretical physicist, born Copenhagen, Denmark, 1885, died Copenhagen, 1962. Nobel Prize for Physics, 1922. Left Denmark in 1943.

Bondi, Hermann, mathematician and cosmologist, born Vienna, Austria, 1919, died Cambridge, UK, 2005. Left Austria in 1937.

Born, Max, mathematical physicist, born Breslau, Germany, 1882, died Göttingen, Germany, 1970. Nobel Prize for Physics, 1954. Left Germany in 1933.

Brauer, Richard, mathematician, born Charlottenburg, 1901, died Belmont, Massachusetts, 1977, left Germany in 1933.

Bretscher, Egon, physicist, born Zurich, Switzerland, 1901, died Switzerland 1973. Left Switzerland in 1936.

Chain, Ernst, biochemist, born Berlin, 1906, died Castlebar, Ireland, 1979. Nobel Prize for Physiology and Medicine, 1945. Left Germany in 1933.

Courant, Richard, mathematician, born Lublinitz, Germany, 1888, died New York, 1972. Left Germany in 1933.

De Benedetti, Sergio, physicist, born Florence 1912, died Sarasota, Florida, 1994. Left Italy in 1939.

Debye, Peter, (1884–1966), physicist and physical chemist, born Maastricht, Holland, 1884, died New York, 1966. Nobel Prize for Chemistry, 1936. Left Germany in 1940.

Dehn, Max, mathematician, born Hamburg, Germany, 1878, died Black Mountain, North Carolina, 1952. Left Norway in 1940.

Delbrück, Max, physicist and molecular biologist, born Berlin, 1906, died Pasadena, California, 1981. Nobel Prize for Physiology and Medicine, 1969. Left Germany in 1937.

Einstein, Albert, mathematical physicist, born Ulm, Germany, 1879, died Princeton, New Jersey, 1955, Nobel Prize for Physics, 1921. Left Germany in 1932.

Eisenschitz, Robert, Austrian physicist, born Vienna, 1898, died London, 1968. Left Germany in 1933.

Elsasser, Walter, German physicist, born Mannheim, 1904, died Baltimore, 1991 . Left Germany in 1933.

Erdös, Paul, mathematician, born Budapest, Hungary, 1913, died Warsaw, Poland, 1996. Left Hungary in 1934.

Fano, Ugo, physicist, born Turin, Italy, 1912, died Chicago, Illinois, 2001. Left Italy in 1939.

Fayans, Kasimir, chemist, born Warsaw 1887, died Ann Arbor, Michigan, 1995, left Germany in 1935.

Fermi, Enrico, physicist, born Rome, Italy, 1901, died Chicago, Illinois, 1954. Nobel Prize for Physics 1938. Left Italy in 1938.

Franck, James, physicist, (1882–1964), born Hamburg, Germany, 1882, died Göttingen, Germany, 1964. Nobel Prize for Physics, 1925. Left Germany in 1933.

Freud, Sigmund, psychoanalyst, born Pribor, Moravia, 1856, died London, UK, 1939. Left Austria in 1938.

Frisch, Otto, nuclear physicist, born Vienna, Austria, 1904, died Cambridge, UK, 1979. Left Germany in 1934.

Fröhlich, Herbert, German physicist, born Rexingen 1905, died Liverpool, 1991. Left Germany in 1933.

Fubini, Eugenio, Italian mathematician, one-time vice-president of IBM, born Turin, 1913, died Brookline, MA, 1997. Left Italy in 1939.

Fubini, Guido, Italian mathematician, born Venice 1879, died New York, 1943, Left Italy in 1939.

Fuchs, Klaus, physicist and spy, born Russelsheim, Germany, 1911, died Dresden, East Germany, 1988. Left Germany in 1933.

Gabor, Denis, electrical engineer, born Budapest, Hungary, 1900, died London, UK, 1979. Nobel Prize for Physics 1971. Left Germany in 1933.

Gamov, George, physicist, born Odessa, Russia, 1904, died Boulder, Colorado, 1968. Left the Soviet Union in 1933.

Gödel, Kurt, mathematician, born Brno, Moravia, 1906, died Princeton, New Jersey, 1978. Left Austria in 1939.

Gold, Thomas, Austrian astrophysicist, born Vienna, 1920, died Ithaca, New York, 2004. Left Austria in 1939.

Goldhaber, Gerson, physicist, born Chemnitz 1924, died Berkeley, California, 2010. Left Germany in 1933.

Goldhaber, Maurice, physicist, born Lviv, 1911, died New York, 2011. Left Germany in 1933.

Gordon, Walter, physicist, born Apolda, 1893, died Stockholm, 1939. Left Germany in 1933.

Haber, Fritz, chemist, born Breslau, Germany, 1968, died Basel, Switzerland, 1934. Nobel Prize for Chemistry, 1919. Left Germany in 1933.

Heitler, Walter, theoretical physicist, born Karlsruhe, Germany, 1904, died Zollikon, Switzerland, 1981. Left Germany in 1933.

Hertz, Gustav, physicist, born Hamburg, Germany, 1887, died Berlin, 1975. Nobel Prize for Physics, 1925. Left Germany in 1945.

Herzberg, Gerhard, spectroscopist, born Hamburg, Germany, 1904, died Ottawa, Canada, 1999. Nobel Prize for Chemistry 1971. Left Germany in 1935.

Hess, Victor, physicist, born Schloss Waldstein, Austria, 1883, died Mount Vernon, New York, 1964. Nobel Prize for Physics, 1936. Left Austria in 1938.

Houtermans, Friedrich, physicist, born Danzig, Prussia, 1903, died Switzerland, 1966. Left Germany in 1933.

Katz, Bernard, biophysicist, born Leipzig, Germany, 1911, died London, UK, 2003. Nobel Prize for Physiology and Medicine 1970. Left Germany in 1935.

Kemmer, Nicholas, theoretical physicist, born St Petersburg, Russia, 1911, died London, UK, 1999. Left Switzerland in 1936.

Kohn, Walter, chemist, born Vienna, Austria, 1923. Nobel Prize for Chemistry, 1998. Left Austria in 1938.

Kowarski, Lew, physicist, born St Petersburg, Russia, 1907, died Geneva, Switzerland, 1979. Left France in 1940.

Krebs, Hans, physiologist, born Hildesheim, Germany, 1900, died Oxford, UK, 1981, Nobel Prize for Physiology and Medicine, 1953. Left Germany in 1933.

Kurti, Nicholas, physicist, born Budapest, Hungary, died Oxford, UK, 1998. Left Germany in 1933.

Lemberg, Max, biochemist, born Beslau 1896, died Sydney, Australia 1975. Left Germany in 1933.

Levi, Giuseppe, neurologist, born Trieste 1872, died Turin, 1965. Left Italy in 1938.

Levi-Montalcini, Rita, neurologist, born Turin, Italy, 1909. Nobel Prize for Physiology and Medicine, 1986, ousted from university 1939.

Levi-Vita, Tullio, mathematician, born Padua, Italy, 1873, died in obscurity in Rome in 1941 after being stripped of authority.

Lipmann, Fritz, (1899–1975), biochemist, born Königsberg, Germany, 1899, died Poughkeepsie, New York, 1975. Nobel Prize for Physiology and Medicine 1954. Left Germany in 1939.

Loewi, Otto, physiologist, born Frankfurt, Germany, 1873, died New York, 1961. Nobel Prize for Physiology and Medicine, 1936. Left Austria in 1938.

London, Fritz, physicist, born Breslau, Germany, 1900, died Durham, NC, 1945. Left Germany in 1933.

London, Heinz, physicist, born Bonn, Germany, 1907, died Oxford, UK, 1970. Left Germany in 1933.

Luria, Salvador, microbiologist, born Turin, Italy, 1912, died Lexington, Massachusetts, 1992. Nobel Prize for Physiology and Medicine 1969. Left Italy in 1938.

Meissner, Karl, physicist, born Reutingen 1891, died on board a ship, 1959. Left Germany in 1938.

Meitner, Lise, physicist, born Vienna, Austria, 1878, died Cambridge, UK, 1968. Left Germany in 1938.

Mendelssohn, Kurt, physicist, born Berlin 1906, died Oxford, 1980. Left Germany in 1933.

Meyerhof, Otto, biochemist, born Hanover, Germany,1884, died Philadelphia, Pennsylvania, 1951. Nobel Prize for Physiology and Medicine, 1922. Left Germany in 1938.

Mortara, Nella, physicist, born Pisa 1893, died Rome 1988. Left Italy in 1939.

Neuberg, Carl, biochemist, born Hannover, 1877. Died New York 1956. Left Germany in 1934.

Noether, Emmy, mathematician, born Erlangen, Germany, 1882, died Bryn Mawr, Pennsylvania, 1935. Left Germany in 1933.

Nordheim, Lothar, physicist, born Munich 1896, died La Jolla, California, 1985. Left Germany in 1934.

Pauli, Wolfgang, mathematical physicist, born Vienna, Austria, 1900, died Zurich, Switzerland, 1955. Nobel Prize for Physics 1945. Left Switzerland in 1940.

Peierls, Rudolph, physicist, born Berlin, Germany, 1907, died Oxford, UK, 1995. Left Germany in 1933.

Penzias, Arno, physicist, born Munich, Germany, 1933. Nobel Prize for Physics, 1978. Left Germany in 1939.

Perutz, Max, molecular biologist, born Vienna, Austria, 1914, died Cambridge, UK, 2002. Nobel Prize for Chemistry, 1962. Left Vienna in 1936.

Placzek, George, physicist, born Brno, Moravia, 1905, died Zurich, Switzerland, 1955. Left Austria in 1932.

Polanyi, Michael, physicist, mathematician, and economist, born Budapest, Hungary, 1891, died Northampton, UK, 1976. Left Germany in 1933.

Pontecorvo, Bruno, physicist, born Pisa, Italy, 1913, died Dubna, Russia, 1993. Left Italy in 1936.

Prager, William, mathematician, born Karlsruhe 1903, died Switzerland, 1980. Left Germany in 1934.

Przibram, Karel, physicist, born Vienna, 1878, died 1973. Left Austria in 1939.

Rabinowitch, Eugene, physicist, born St Petersburg, Russia, 1901, died Washington, DC, 1973. Left Germany in 1933.

Racah, Giulio, mathematician, born Florence, Italy, 1909, died Florence, 1965. Left Italy in 1939.

Rasetti, Franco, physicist and botanist, born Castiglione, Italy, 1901, died Waremme, Belgium, 2001. Left Italy in 1939.

Riefenstahl, Charlotte, physicist, born Bielefeld, 1899, died 1978. Left Germany in 1933.

Rossi, Bruno, physicist, born Venice, Italy, 1903, died Cambridge, Massachusetts, 1993. Left Italy in 1938.

Rotblat, Josef, physicist, born Warsaw, Poland, 1908, died London, UK, 2005. Nobel Peace Prize 1995. Left Poland in 1939.

Rothe, Erich, mathematician, born Berlin 1895, died 1988. Left Germany in 1937.

Scharff-Goldhaber, Gertrud, physicist, born Mannheim, 1911, died Patchogue, NY, 1998. Left Germany in 1935.

Schrödinger, Erwin, mathematical physicist, born Vienna, Austria, 1887, died Vienna, 1961. Nobel Prize for Physics 1933. Left Germany in 1933.

Segrè, Emilio, physicist, born Tivoli, Italy, 1905, died Lafayette, Indiana, 1989. Nobel Prize for Physics, 1959. Left Italy in 1938.

Simon, Francis, physicist, born Berlin, 1893, died Oxford, UK, 1956. Left Germany in 1933.

Smoluchowski, Roman, Polish physicist, born Zakopane, Austria-Hungary, 1910, died Austin, TX, 1995. Left Poland in 1939.

Steinberger, Jack, physicist, born Bad Kissingen, Germany, 1921. Nobel Prize for Physics, 1988. Left Germany in 1934.

Stern, Otto, atomic physicist, born 1888, Sohrau, Prussia, died Berkeley, California, 1969. Nobel Prize for Physics 1943. Left Germany in 1933,

Szego, Gabor, mathematician, born Kunhegyes, Hungary, 1895, died Palo Alto, California, 1985. Left Germany in 1934.

Szilard, Leo, nuclear physicist, microbiologist, and political agitator, born Budapest, Hungary, 1898, died La Jolla, California, 1964. Left Germany in 1933.

Teller, Edward, nuclear physicist and protagonist, born Budapest, Hungary, 1908, died Stanford, California, 2003. Left Germany in 1933.

Ulam, Stanislaw, mathematician and physicist, born Lvov, Poland, 1909, died Santa Fe, New Mexico, 1984. Left Poland in 1938.

Von Halban, Hans, physicist, born Leipzig, Germany, 1906, died Paris, 1964. Left France in 1940.

Von Hevesy, Georg, chemist, born Budapest 1885, died Freiburg, Germany, 1966. Nobel Prize for Chemistry 1943. Left Denmark in 1943.

Von Hippel, Arthur, physicist, born Rostock, 1898, died Boston, Massachusetts, left Germany in 1934.

Von Karman, Theodor, physicist and aeronautics engineer, born Budapest 1881, died Aachen, Germany, 1963. Left Germany in 1930.

Von Mises, Richard, mathematician, born Lvov, 1883, died Boston, Massachusetts, 1953. Left Germany in 1933.

Von Neumann, John, mathematician, born Budapest, Hungary, 1903, died Washington DC, 1955. Left Germany in 1930.

Weill, Kurt, composer, born Dessau, 1900, died, New York, 1990. Left Germany in 1933.

Weisskopf, Victor, nuclear physicist, born Vienna, Austria, 1908, died Newton, Massachusetts, 2002. Left Germany in 1933.

Weyl, Hermann, mathematician, born Elmshorn, Germany, 1885, died Zurich, Switzerland, 1955. Left Germany in 1933.

Wigner, Eugene, mathematical and nuclear physicist, born Budapest, 1902, died Princeton, New Jersey, 1995, Nobel Prize for Physics 1963. Left Germany in 1930.

Willstätter, Richard, biochemist, born Karlsruhe, Germany, 1872, died Locarno, Switzerland, 1942. Nobel Prize for Chemistry, 1915. Left Germany in 1938.

Wittgenstein, Ludwig, philosopher, born Vienna, 1889, died Cambridge, UK, 1951. Left Austria in 1929.

Appendix 2: Sources and bibliography

The Nazi persecution of the Jews and the development of the Atomic Bomb are among the most well-documented events in contemporary history. For these two topics, I relied mainly on the authoritative works of Saul Friedländer and Richard Rhodes, respectively. Anyone wishing to learn more should first read these books. Together with the other main sources referred to in the footnotes, they are listed here.

However other books, websites, archived papers, articles in learned journals, and more obscure items are indicated explicitly in the footnotes. There are many published biographies and autobiographies of the leading scientists of the *Quantum Exodus*. Supplementing these books are the Biographical Memoirs (obituaries) of the Fellows of the Royal Society, London, and the oral interviews held by the Niels Bohr Library and Archives of the American Physical Society (Niels Bohr is an honorific title: the archives cover hundreds of scientists). While I have made free use of this material, many of these books and other documents make little reference to their subjects' Jewishness, nor to the trauma of having to flee from Nazi Europe. The anguish and tribulation came more from the experience of ordinary people who could not achieve such elevated academic status.

BOOKS

In the footnotes, references to these books are indicated simply by the author's name, e.g. Beyerchen, p.56.

Adam, U., *Judenpolitik im Dritten Reich*, Droste, Düsseldorf, 2003.

Amaldi, E., Battimelli, G., and Paoloni, G., *20th Century Physics: Essays and Recollections* World Scientific, Singapore, 1998.

Arms, N., *A Prophet in 2 Countries, the Life of F.E. Simon*, Pergamon, Oxford, 1966.

Bernstein, J., *Hans Bethe, Prophet of Energy*, Basic Books, Cambridge, MA, 1980.

Bernstein, J., *Plutonium*, Joseph Henry Press, Washington, DC, 2007.

Beyerchen, A., *Scientists Under Hitler – Politics and the Physics Community in the Third Reich*, University of Yale Press, New Haven, 1977.

Born, M., *My Life*, Taylor and Francis, London, 1978.

Brustein, W., *Roots of Hate, Anti-Semitism in Europe before the Holocaust*, Cambridge University Press, Cambridge, 2003.

Byers, N., (ed) *Out of the Shadows, Contributions of Twentieth-Century Women to Physics*, Cambridge University Press, Cambridge, 2006.

Bryson, B., *A Short History of Nearly Everything*, Doubleday, New York, 2003.

Campbell, J., *Rutherford, Scientist Supreme*, AAS Publications, Christchurch, NZ, 1997.

Cassidy, D., *Uncertainty, The Life and Science of Werner Heisenberg*, W H Freeman, Basingstoke, 1991.

Charpa, U. and Deichmann, U. (eds), *Jews and Sciences in a German Context*, Mohr Siebeck, Tübingen, 2007.

Charpak, G., *La Vie à Fil Tendu*, Odile Jacob, Paris, 1993.

Cornwell, J., *Hitler's Scientists: Science, War and the Devil's Pact*, Penguin, London, 2004.

Dahl, P., *Nuclear Transmutation to Nuclear Fission, 1932-1939*, Institute of Physics Publishing, Bristol, 2002.

Elon, A., *The Pity of it All, A Portrait of the Jews in Germany, 1743-1933*, Penguin, London, 2002.

Engel, D., *The Holocaust: The Third Reich and the Jews*, Longman, London, 2000. (This short but authoritative book includes a full bibliography on this subject.).

Enz, C., *No Time to be Brief, A Scientific Biography of Wolfgang Pauli*, Oxford University Press, Oxford, 2010.

Evans, R., *The Coming of the Third Reich, How the Nazi Destroyed Democracy and Seized Power in Germany*, Penguin, London, 2004.

Farmelo, G., *The Strangest Man, The Hidden Life of Paul Dirac, Quantum Genius*, Faber, London, 2009.

Fermi, L., *Atoms in the Family, My Life with Enrico Fermi*, University of Chicago Press, 1954.

Friedländer, S., *Nazi Germany and the Jews*, Wiedenfeld and Nicholson, London, 1997, subsequently reissued as two volumes, the first being *The Years of Persecution, Nazi Germany and the Jews 1933-39*, Phoenix, London, 2003.

Frisch, O., *What Little I Remember*, Cambridge University Press, Cambridge, 1991.

Gilbert, M., *First World War*, Wiedenfeld and Nicholson, London, 1994.

Gowing, M., *Britain and Atomic Energy, 1939-1945*, Macmillan, London, 1964.

Greenspan, N., *The End of the Certain World, The Life and Science of Max Born*, Basic Books, New York, NY, 2005.

Hargittai, I., *Martians of Science, Five Physicists who Changed the Twentieth Century*, Oxford University Press, Oxford, 2008.

Heilbron, J., *Max Planck, Dilemmas of an Upright Man*, U. California Press, 1986.

Heim, S., Sachse, C., and Walker, M. (eds) *The Kaiser Wilhelm Society under National Socialism*, Cambridge University Press, Cambridge, 2002.

Hertz, D., *How Jews Became Germans*, Yale University Press, 2007.

James, I., *Driven to Innovate – A Century of Jewish Mathematicians and Physicists*, Peter Lang, Oxford, 2009.

Johnson, P., *History of the Jews*, Phoenix, London, 1993.

Jungk, R., *Brighter than 1000 Suns*, Penguin, London, 1956.

Lanouette, W., *Genius in the Shadows, Biography of Leo Szilard, the Man Behind the Bomb*, Scribner's, New York, NY, 1992.

Lanzmann, C., *Le Lièvre de Patagonie, Mémoires*, Gallimard, Paris, 2009.

Levi, P., *The Periodic Table*, Penguin, London, 1975.

Lindemann, A., *Esau's Tears: Modern Anti-Semitism and the Rise of the Jews*, Cambridge University Press, Cambridge, 1997.

Longerich, P., *Davon haben wir nichts gewusst! Die Deutschen und die Judenverfolgung, 1933-1945*, Pantheon, München, 2007.

McDonough, F., *Hitler & Nazi Germany, Perspectives in History*, Cambridge University Press, Cambridge, 1999.

Macrae, N., *John von Neumann*, Pantheon, New York, 1992.

Medawar, J. and Pyke, D., *Hitler's Gift –Scientists who Fled Nazi Germany*, Piatkus, London, 2000.

Mendelssohn, K., *The World of Walther Nernst: The Rise and Fall of German Science*, Macmillan, London, 1973.

Moore, W., *Schrödinger, Life and Thought*, Cambridge University Press, Cambridge, 1989.

Niederland, D., *Jewish Emigration from Germany in the First Years of Nazi Rule*, Leo Baeck Institute Yearbook, New York, NY, 1998, p.285.

Noakes, J. and Pridham, G., *Nazism 1919-1945* Vol 2, Exeter U. Press (2002).

Pais, A., *Subtle is the Lord, The Science and the Life of Albert Einstein*, Oxford University Press, Oxford, 1982 (referred to as Pais SL).

Pais, A., *Inward Bound, On Matter and Forces in the Physical World*, OUP, Oxford, 1986 (Pais IB).

Pais, A., *Niels Bohr's Times, In Physics, Philosophy and Polity*, Oxford University Press, Oxford, 1991 (Pais NBT).

Pais, A., *Einstein Lived Here*, Oxford University Press, Oxford, 1994 (Pais ELH).

Peierls, R., *Bird of Passage, Recollections of a Physicist*, Princeton University Press, 1985.

Poundstone, W., *Prisoner's Dilemma, John von Neumann, Game Theory, and the Puzzle of the Bomb*, Oxford University Press, Oxford, 1992.

Rhodes, R., *The Making of the Atomic Bomb*, Touchstone, New York, NY, 1986 (referred to as Rhodes MAB).

Rhodes, R., *Dark Sun – The Making of the Hydrogen Bomb*, Simon and Schuster, New York, NY, 1995 (Rhodes DS).

Roth, C., *The Jewish Contribution to Civilization*, Macmillan, London, 1938 (referred to as Roth JCC).

Roth, C., *A Short History of the Jewish People*, Hartmore House, New York, NY, 1969 (Roth SHJP).

Rife, P., *Lise Meitner and the Dawn of the Nuclear Age*, Birkhäuser, Bostson, MA, 1999.

Segrè, E., *Enrico Fermi, Physicist*, University of Chicago Press, 1970.

Sime, R., *Lise Meitner, A Life in Physics*, Berkeley UCP, 1996.

Steinberger, J., *Learning about Particles*, Springer, Berlin, 2004.

Strauss, H. (ed), *Jewish Immigrants of the Nazi Period in the USA*, Vol 4 Jewish Emigration from Germany, 1933–42, Sauer, Detroit, 1992 (referred to as Strauss, Immigrants).

Strauss, H., ed, *Emigration. Deutsche Wissenschaftler nach 1933. Entlassung und Vertreibung,* Technische Universität, Berlin, 1987. (referred to as Strauss, Emigration).

Swinne, E., *Richard Gans, Hochschulleher in Deutschland und Argentinien*, ERS, Berlin, 1992.

Szanton, A., *The Recollections of Eugene P Wigner*, Basic Books, Cambridge, MA, 1992.

Teller, E. and Schoolery, J., *Memoirs, A Twentieth Century Journey in Science and Politics*, Perseus, Cambridge MA, 2001.

Ulam, S., *Adventures of a Mathematician*, Scribner's, New York, NY, 1976.

Underwood, M., *Joseph Rotblat, A Man of Conscience in the Nuclear Age*, University of Sussex Press, Brighton, 2009.

Wang, J., *American Science in an Age of Anxiety: Scientists, Anticommunism and the Cold War*, University of North Carolina Press, Chapel Hill, NC, 1999.

Waugh, A., *The House of Wittgenstein, A Family at War*, Bloomsbury, London, 2008.

Weisskopf, V., *The Joy of Insight, The Passions of a Physicist*, Basic Books, Cambridge, MA, 1992.

Williams, R., *Klaus Fuchs, Atom Spy*, Harvard University Press, 1987.

Archive sources

AIP – American Institute of Physics; oral history interviews held by the Niels Bohr Library and Archives, http://www.aip.org/history/nbl/ohilist.html

SPSL – Society for the Protection of Science and Learning, Bodleian Library Oxford, Special Collections,
http://www.bodley.ox.ac.uk/dept/scwmss/wmss/online/modern/spsl/spsl.html

WL – Wiener Library Institute of Contemporary History, London, http://www.wienerlibrary.co.uk/

Acknowledgements

For helping to shape the book, I would like to thank my commissioning editors, Sonke Adlung, Clare Charles, and April Warman at Oxford University Press, and Peter Tallack of the Science Factory. Many thousands of words were written in draft before the book even began to take its present form. A strong influence was Professor Michael Berkowitz at the Department of Hebrew and Jewish Studies of London's University College, who also pointed me towards valuable archive sources. Thanks go to Alexandra Ben-Yehuda at the Liberal Jewish Synagogue, London; Luisa Bonolis in Rome for helpful source material on Mussolini's race laws, and on some of the German emigrants; Monika Wilson for help with translation of items printed in impenetrable Gothic typefaces; the CERN library, Geneva, for its efficient external loans service; and Alison Goddard at the *Economist*, London, for some useful hints. Herwig Schopper, who claims to be the last living scientist to have worked with Lise Meitner, remembered her and Otto Frisch. Writing a book like this means tramping around in long grass which obscures the view for an unwary explorer. I am therefore very grateful to Andy Pearce of London's Royal Holloway College for reading draft material on the Weimar Republic and Nazi Germany. I am especially indebted to Jack Steinberger, 1988 Nobel Physics laureate and himself an emigrant from 1930s Germany, for recollections of his early contemporaries, and for his valuable suggestions after kindly reading the entire draft text.

For help with archives, whether through correspondence or actual visits, I appreciate the help of Howard Falksohn and the efficient staff of the Wiener Library, London. As well as providing a mass of valuable material, the Wiener Library has done a magnificent job in cataloguing and cross-referencing its information—especially difficult when much of this source material is in German—and making it accessible on-line. Svetlana Burmistr at the *Zentrum für Antisemismusforschung, Technische Universität*, Berlin and Adi Fokshenianu at CARA (Council for Assisting Refugee Academics), London were also helpful. It is a special pleasure to thank Colin Harris and the staff of the Special Collections at the Bodleian Library, Oxford. Another invaluable source was the compilation of oral histories in the Niels Bohr Library and Archives of the American Physical Society. The transcripts have the additional advantage of being available on-line. However this precious material naturally concentrates on science issues rather than personal history.

I am especially grateful to Sonke Adlung and to the Weiner Library who took the trouble to find experts to read specific chapters. It is difficult to write what is essentially a background sketch in a way which satisfies such authorities, and I thank them for their patience and understanding. For help with illustrations, I appreciate the assistance from Kevin Roark of the Communications Office at Los Alamos National Laboratory, and the very efficient service provided by the various photographic agencies addressed, especially the Emilio Segrè Visual Archives of the American Institute of Physics.

With judicious use of Wikipedia and other sources, especially when comparing different language versions, one can sometimes accomplish amazing things within the space of a few clicks. Among the websites trawled, the *Deutsche National Bibliothek,* and the second-hand department of *amazon.de* were especially fruitful. I am thankful to my wife Gill for reading the initial draft text and for enduring the painful gestation of another book, and to our children. Nathalie and Ben, for their help, suggestions, and support. Ben also produced the idea for the cover.

Sarah Stephenson and the skilled production team at Oxford University Press transformed my amorphous pile of text and illustrations into an attractive book.

Gordon Fraser
Divonne-les-Bains
August 2011

Index